Army Talk

LONDON: HUMPHREY MILFORD
OXFORD UNIVERSITY PRESS

Army Talk

A Familiar Dictionary of Soldier Speech

By Elbridge Colby

PRINCETON UNIVERSITY PRESS · PRINCETON

Second Edition, 1943

Copyright, 1942, by Princeton University Press
Printed in the United States of America by
Princeton University Press at Princeton, New Jersey

Illustrations by Richard Hurd

To H. L. MENCKEN

PREFACE

This is an attempt to put in a book the language that lives on the lips of fighting men in the army of the United States.

Such a language naturally changes somewhat with the times. It collects and invents new components, for words are often the reflections of events. As Jespersen pointed out in his *Growth and Structure of the English Language*, "alarm" and "corporal" and "sentinel" reach to the wars between Italy and France, "plunder" was brought in from the German by soldiers who had fought under Gustavus Adolphus, and "loot" was learned from the Hindus by English army men in the eighteenth century. Since armies were already highly organized by the time America acquired an individuality of its own, our American military language derives much directly from the English, but some of it stems from Mexican Border affrays and from the "days of the Empire" in the Philippines. Parts were adopted in France in 1917-1918 from the British (e.g., "tank" and "brass hat") and from the French (e.g., "pursuit plane" and "four-ragere"). A very great deal is native. As official regulations change, words which lean heavily on regulations tend to drop out or to be replaced by substitutes, as "Squads East and West" now slowly disappears, as "P.X." gradually gains the ascendancy over "canteen," and "Section Eight" completely supplants "One-forty-eight-and-a-half." These changes are proofs that it is a living language.

It has been sometimes difficult to decide what to exclude, for there is no intention of preparing a formal and complete military dictionary. Purely technical phrases can be understood only after a study of the mechanism of weapons and the technique of tactics; they are used by soldiers only in purely technical talk and instruction. Neither the public at large nor the scholar is interested in these. A "recoil spring rod" and a "magazine floor plate" do not enter conversation, unless a man is talking shop. Yet, there are many presumably technical phrases and official terms frequently used in common conversation with civilians as well as with other soldiers. Soldiers use them when they talk to you and to me. Such of these as appear in the everyday talk of soldiers, I have saved.

Slang of course has been included. Some of it consists of distortions or abbreviations of official terms, resulting from the common processes of language. Some of it springs out of environment, incident, analogy, or pure imagery by the spontaneous and unpredictable processes of veritable slang. My principal difficulty with this has been to decide what is clearly military in character and what is American generally. For example, take the old phrase "fisheyes," which is used to denote tapioca pudding. Even though used in mess hall, it is hardly army slang; it is boarding-school slang, Boy Scout slang, boys' camp slang, even girls' camp slang—in fact, actually universal "kid" slang. To be army slang, a phrase must either have originated in or be peculiar to the army.

I am reminded of what *The Stars and Stripes* said as long ago as April 12, 1918: "Universal slang in this man's army is as hard to attain as universal peace in this man's world." It is hard, and the process is slow. It is inevitable therefore that there will be difficulty with regard to some recently devised slang terms, invented since the beginning of our mobilization of 1940.

For example, take the new word "jeep." At Fort Knox, we are told, it is used for a "command car," at Fort Meade for the new "bantam" car; at Fort Dix it is synonymous with "recruit." The word is running around and around and is being used so loosely that it will take time, settling, and conventionalizing before it can be said to have any definite meaning. Indeed, the bantam cars themselves are in different places variously called "bantams," "jeeps," "peeps," "blitzbuggies," and even "jaloppies."

Like "jeep," there are many current slang words so recently applied to one or so many different things that it is impossible to record them yet as acknowledged and generally adopted. When sufficient time has passed for particular and popular usages to become settled, some of the needs for this book will have passed. So, ten years hence it may be found that the accidents of popularity have fixed some which I should have included, and have excluded some which I here record, although it should be said that I have tried to be strict in holding up the bars. If this attitude seem to be too conservative, I can only reply that an army is by nature conservative too. Also, it must be remembered that once a word gets established in army speech, it most likely remains. Only a week from the day on which these lines are being written, I heard some newly drafted men use the terms "topkick," "canteen check," and "jawbone"—terms which have been in the army for more than forty years—on the lips of soldiers who had been in less than forty days. In the history of soldier speech, each new generation jauntily adopts and uses the language of the "old soldier" and, even in times of great mobilizations and changes in equipment, the number of new slang terms is always very small.

Whether words be technical jargon familiar in the non-technical conversation of the private soldier, or essentially slang in character, any person interested in language as such or in the soldier as a human being will be interested in origins. These I have given, when found.

[xi]

In order, however, to satisfy the curiosity of those interested in new terms whose use is neither old enough nor wide enough to warrant admission to the canon of established speech, I have given at the end a supplementary list of a few phrases and words likely to be heard here and there, but uncertain as to origin and as to validity. From even this supplementary list I have excluded distinctively West Point slang as foreign to the army as a whole even though occasionally and unusually introduced by young Academy graduates not yet entirely free from schoolboy habits and speech.

So, I have tried to collect such soldier speech of these types as has come my way during more than two decades of military life, to analyze it for its origins and connotations, and to record it here for the amusement as well as for the edification of all who are interested in the army. It is difficult to hope that the collection is complete, even though I gave my early attempts in this direction tentative publication ten years ago in an army periodical, received as a result many corrections and additions, and have worked since that time to increase, perfect, and modernize the collection.

To the editors and publishers of *Our Army*, *The Wight House Press*, *The Military Engineer*, and *American Speech*, I am indebted for many kindnesses as well as for their permission to reproduce what they originally put into print.

I express my thanks to George Philip Krapp who awakened my interest in language; to Henry L. Mencken, who encouraged me in this work; and to Peter M. Miller, Jr., who gave keen assistance. I have of course leaned pretty heavily in spots on the *New English Dictionary*, as anyone doing work of this sort should. Special obligations for permission to quote are listed under "Acknowledgments" in the back of the book.

I am interested in this language as a living thing, a reflection of fact and circumstance, and would be sincerely sorry if this publication should tend to standardize and fix usages and meanings. A language is interesting only as it changes, reflecting the flow of life and the change of pattern which it depicts.

To philosophize upon this language in a brief introduction would be to attempt inadequately what should be a detailed study rather than a preface. It is possible, however, to state generally some broad impressions.

1. Soldier speech tends to abbreviations, sometimes being led in that direction by officialdom.

2. Soldier slang leans toward the uncomplimentary, and avoids the sentimental.

3. Most of our talk comes from the British, much of it also is native in origin, and some is drawn from the French and the Spanish directly.

4. Even in moments of relaxation, troops talk a jargon replete with technical phrases.

5. Army language is highly specialized, the historic origins of very many of the words bearing no relation to their specialized modern meanings.

6. In general an army resists change, and its familiar phrases continue conservatively, handed down to successive generations of soldiers. Few of them drop out.

I am greatly indebted to many readers who have aided me in making this Second Edition somewhat more complete and up-to-date than its predecessor.

E. C.

Army Talk

Army Talk

ACK-ACK. British for antiaircraft fire, taken from the signal corps phonetics for A-A. Popular with Americans in the Pacific. Over Europe "flak" is used more.

A. D. C. This is the aide-de-camp, a title taken directly from the French many years ago meaning literally "camp assistant," and used the world over to designate a junior officer specially detailed to assist a general. Sometimes he serves as military secretary, sometimes as social secretary. Sometimes he looks after the general's field baggage and equipment or buys his railroad tickets. Sometimes he is a fixture at tea parties given by the general's wife, and runs with her letters to the post office. In Washington there are White House "aides" who accompany the President himself, and "junior aides" at the White House who must dress up in rich blue and gleaming gold and form lines like ushers or traffic police as people pour in for formal receptions. The aide has his privileges. In 1808, Wellington said: "That spare room may be kept for my horses and those of my Aides-de-Camp." In those days, aides were frequently used in carrying important messages on the battle field. Napoleon had many such. The word is usually abbreviated in speaking to "aide." According to *Punch*, the British soldier calls him the *'An de cap* to a general; that is, the one who hands the general his cap and performs such services, or perhaps derisively just a *handicap*. In the army, aide is an official title, and is never used, as it is on the outside, in a general sense to refer to subordinates or supporters, as "political aides." The spelling has not been anglicized, although the word has, and although the "aide" is frequently—and inaccurately—written without the "e" and so seen in America from the eighteenth right through to the twentieth century.

ADJUTANT. In general, an assistant or a helper. Said by the *New English Dictionary* to be in a military sense "an officer in the army whose business it is to assist the superior officers by receiving and communicating orders, conducting correspondence, and the like." A battalion adjutant is usually a lieutenant to assist the lieutenant colonel in command, and a regimental adjutant is usually a captain or a major to assist his colonel in all administrative routine. These do the office work while others drill the troops. They are most conspicuous out of doors when a formal "parade" or "review" is held, when they stand at the flank of the line, then march alone out in front while everyone else stands rigid and waits and wishes they would walk faster, and finally call for reports and tell the Old Man that all are "present or accounted for." There are also adjutants for posts, camps, forts, and other stations. Livy speaks of Roman times and the "adjutants whom our ancestors thought to give their generals in war." If all generals wrote home only three-word reports of campaigns like Caesar (*Veni, vidi, vici!*), adjutants would not have much to do. But now every headquarters has at least one mimeograph machine and adjutants are kept busy. In the United States, the word "adjutant" has taken the place of an earlier "military secretary." The adjutant is the administrative officer. He sees that the strength reports are properly made up; he prepares the rosters of strength and looks after the pay rolls. When the commander has decided what he wants done and how, the adjutant issues the necessary order or orders and signs, putting above his own

[4]

signature the words: "By command of Brigadier General Blank" or "By order of Colonel Blunk."

ADJUTANT'S CALL. A special bugle call of the army, used prior to certain specified ceremonies. Usually before a formation, the bugles play "First Call" as a warning, and then the "Assembly." However, prior to guard mount, and formal parades and reviews, there is played "Adjutant's Call," for on those occasions the units called out have to be reported to the adjutant. These words have been written:

> Winter or spring, summer or fall,
>   This is the sound of Adjutant's Call.
> Winter or spring, summer or fall,
>   This is the sound of Adjutant's Call.

ADJUTANT GENERAL. Originally (1645) the title of an adjutant assigned to "assist the general of an Army." The words are also used officially for a special department of the army composed of specially trained and experienced administrative officers, called the Adjutant General's Department, who provide adjutants for important posts, for divisions, corps, and higher units. Such officers are "Adjutant Generals." There are lots of them. They wear little three colored United States shields on the lapels for insignia. In the American Army, "The" Adjutant General is a sacred title, held only by the chief of that department who sits in the circles of the mighty in Washington, D.C., and among other duties gets out the orders which are issued "By order of the Secretary of War."

[5]

AIGUILLETTE. An ornamental tagged or pointed cord or braid worn sometimes on the uniform over the left shoulder—that is you wear it if you are entitled to it. Many units of the First Division and the Second Division in the United States Army wear a cloth aiguillette in the red and green colors of the Croix de Guerre, awarded these units by the French government and called the fourragere. Larousse, that compendium of all French information, says that this decoration is a representation of the hangman's noose, worn by one-time cowardly Flemings under the Duke of Alva who later so distinguished themselves that their distinctive badge of shame became a mark of glory and it was adopted as a decoration by crack regiments. With the dress uniform a different kind of aiguillette made of twisted gold cord is worn by aides and adjutants. In England you will find them called the "aiglets" of a staff captain, or of a dragoon guards officer all dressed up, and long ago spelled "aglet." These, you see, go with a job. In ancient years, before metal armor was invented, the knights of old wore leather body covering which had to be laced up the back. Their squires were required to carry long leather thongs, pointed with bone, ready to lace up their knights. Squires were field assistants to their knights. So the modern assistants inherit the badge of service.

APPROACH MARCH. This is a technical army term, still so used, and also perverted and transformed into sly humor. Strictly speaking, the approach march takes place on campaign. Of course, when you are marching you are approaching something, unless you are marching too long in the same place. In this case it is the enemy you are approaching, and the technicians in military phraseology tell us that the "approach march" is that part which extends from the place and time when the enemy shells or fire begin to come down on you, until you get close enough to fire at him with your own rifles. By analogy, when the buck private gets called to the orderly room for an interview with the Top Kick or the Old Man, he slicks his hair and straightens his blouse, and advances with trepidation. He calls this his "approach march." Informally it is sometimes shouted as the equivalent of "come here" or "front and center." In other branches than the infantry it has another name—in the Air Corps it is "taxi up" and in many motor outfits it is "drive up."

ARCHIES. Shells from artillery exploding in the air when fired at aircraft. The puffs of smoke around flying planes during the World War were rarely evidence of more than inaccurate shooting. The nickname came out of the British Army and was generally adopted by Americans. It is supposed to have been a carryover from a line of a popular song of that time which said: "Archibald, certainly not!" Antiaircraft fire has in the two intervening decades become much more accurate and effective; but the puffs of smoke from exploding antiaircraft shell are still called "archies." Indeed, the antiaircraft guns themselves are familiarly called "archies" too. We have begun to use "Ack-Ack" for this, and the "flak" borrowed from the Germans.

ARMORED COW. The old phrase "canned cow" for condensed milk has in the modern mechanized army, especially in the Armored Forces, been changed to "armored cow" and so spread pretty generally throughout the service.

ARMORY. A place where arms and armor are kept. Notice that in *Titus Andronicus* Shakespeare speaks of "the goodliest weapons of his armorie." In the American service the term is used almost exclusively to refer to the drill hall

[7]

or building of a National Guard unit; and this is strictly correct, for our National Guard or State Guards do not take their weapons to their homes as the Swiss do. They are kept in the armory and there secured for the drill periods. The meaning has been extended, however, so as to refer to the building as a whole and not merely to the particular storeroom for the rifles and guns. In addition there is an occasional use of the word in America to refer to a place where arms are manufactured. "Arsenal" for military and "gun foundry" for civilian establishments are widely used and generally understood now in this sense. But it was not always thus. In the early days of our history an arsenal was the storage place; today it is the manufacturing place, like Rock Island Arsenal, Frankfort Arsenal, Watertown Arsenal. The change had official sanction when Congress in 1794 "established at each of the aforesaid arsenals, a national armory"—using the two words in the opposite sense from what we use them today. The manufacturing sense of the word "armory" clings only in the official title of the Springfield Armory; all other armories, as we have said, are storage places where drills are also held.

ARMY. An armed force. The word comes from the Latin *armata* through the French *armée*. Kames (1762) says: "A number of men under the same military command, are termed an army." There is a special use of the term "army." When two or more divisions are grouped under a single commander, the group is called a corps. Two or more corps grouped together under a single command will take the title army. This use of these terms was common in the days of Napoleon, when he had "The Army of Italy" and the "Army of the Sambre and Meuse," and we find it in the American Civil War in "The Army of the Potomac" and "The Army of the Cumberland" and the rest, not to men-

[8]

tion "The Army of Northern Virginia" which never got south of Richmond until it was almost ready to surrender. Since troops do not necessarily remain in one geographical locality, it was perhaps inevitable that the geographical designations should give way to mere numbers. For example, in the World War, our earliest divisions in France were formed into corps; then these corps were grouped into the First Army for the battle of St. Mihiel. During the Argonne battle there were so many corps that it was found better to group them separately, and so some of them were made into a Second Army. When more were marched into Germany, they were labelled "The Third Army"—and it was these who arrived at Coblenz and were not called "The Army of the Rhine" but rather the Third Army. When such an army (or field army, if you will) is formed, it needs supply and communication troops additional to those in the corps under it; such troops and units are spoken of as "Army Troops" and work directly under control of army headquarters, not under the corps and divisions which form the army, which have "corps troops" and "divisional troops" of their own.

ARMY BRAT. Children of army officers have proverbially been almost as undisciplined as army wives. They range about the drill grounds, through the Post Exchange, and among the barracks as if they owned the entire reservation. Among enlisted men, they quickly acquired the name "army brats" which was taken up in semi-seriousness even by army parents and by the "brats" themselves.

ARREST. In civil life, arrest means to take into custody. It means the policeman, sheriff, deputy, or constable seizing you and taking you to jail. In the army, for minor offenses, it is not necessary to put a man in the guard house (or army jail), but simply to put him in "arrest" in the army way. In the army, "arrest" means that you are relieved from duty and must remain in your quarters—or other broader limits may be specifically assigned—until the proper authorities have investigated your case or you have been tried by court-martial. Indeed the army does not use the word "arrest" as a verb at all, but uses it only as a noun, speaking of being "in arrest" or being "put in arrest." When you go to the guardhouse, you are not "arrested"; you are "confined."

ARTICLES OF WAR. These "Articles of War" are the basic law of the army enacted by Congress. They reach back in history to Henry V and to Gustavus Adolphus who devised rules for the better government of their armies, rules fitted for military circumstances for which civil law was not designed and was not appropriate. In 1776 the American colonies adopted articles imitated in large measure from the British. They have been several times revised and perfected to fit the temper of the American people and the changing circumstances of the times. So that all will have notice of their being "subject to military law," it is required that the Articles of War be read and explained to all recruits soon after enlistment, and that this reading be repeated at intervals throughout a soldier's service. The Articles authorize courts-martial and procedure, and give specific jurisdiction of offenses, like absence without leave, desertion, insubordination, and the like. Then there is one which says that "although not mentioned in these articles, all . . . disorders and neglects to the prejudice of good order and military discipline, all conduct of a nature to bring discredit upon the military service, . . . shall be taken cognizance of by a summary, special or general court-martial, and punished according to the nature and degree of the offense."

Artillery

ARTILLERY. A word borrowed from the French and now adapted to more modern weapons. It was once used to indicate any contrivance for throwing missiles—catapults, slings, bows, and guns. For example the famous "Honorable Artillery Company of London" was actually a company of archers, who were authorized to organize and to meet and practise shooting arrows on the "Artillery" ground in London. Now the word has come to mean large guns, and that branch of the service which operates those guns. Notice how Daniel in his *Civil Wars* (1597) used this new meaning in the age of Shakespeare: "Artillerie, th' infernall instrument, New brought from hell to scourge mortalitie, With hideous roaring, and astonishment." At first all artillery was heavy and slow to move around. Frederick the Great was the first to give his guns great mobility and to use them on field campaigns, as well as at sieges, shoving them right up in the front lines with his infantry. Napoleon, an artillerist by training, continued the practice. In England after a while they began, in our language, to distinguish between the heavy and immobile "fortress" artillery and the lighter and mobile "field" artillery. In the United States in 1907, the Field Artillery was completely divorced from the Coast Artillery on the same sort of distinction, although this distinction has somewhat broken down with the passing of time.

ARTILLERY BULL. A complete miss on the target range. When the target has not been struck at all and the firer cannot even get two points for his shot, but gets just a plain zero, a flag is waved across the face of the target. Because this flag is of the red color which is used on guidons, standards, and trouser stripes in the artillery branch of the service, such a shot is called "an artillery bull's eye." Frequently you will find the flag referred to as "lingerie"—which is a slang adaptation of the red petticoat idea—but don't be surprised at that, for the soldier is always thinking up new extensions of his familiar phrases. Indeed, this discouraging red flag is also called "Aunt Nancy's pants," "Aunt Maggie's drawers," and "Maggie's drawers"—the last possibly most frequently.

AS YOU WERE. A military phrase of command in the drill regulations of both the British and American armies which rescinds a previous command. If you make a mistake and desire to start all over again, you use this phrase. For instance: "Right . . . AS YOU WERE . . . Left . . . FACE!" Of course, if the men in ranks start to do something wrong, the commander often has to use "As you were" to unsnarl them and get them back the way they were so he can start them all over again. There is another use of the phrase. An officer comes by; men stand to attention and start to salute; he says: "As you were!" and all formalities are off. The plural pronoun "you" which might have been taken personally, has become completely impersonal and a mere part of a phrase.

ASSEMBLY. This word, very common in civil life, is practically never used in the army, except to refer to the bugle call of that name, which is blown at the last instant that a formation is ordered, just as the first sergeant's voice booms out: "Fall In!"

Moderato.

AT EASE. This is a word of command. Before the World War, when troops were standing in ranks at the rigid position of attention, this command permitted them to relax, so long as they retained one foot in place and remained silent. Then, during that conflict, American forces began to imitate the "Stand at ease" of the British, a command which moved the left foot a prescribed distance to the left, leaned the rifle in the right hand out rigidly to the right front, and shot the left forearm behind the back. This "Stand at ease" position got into our drill books after the war, and remained for two decades. Recent changes in the regulations have eliminated the command "Stand at ease" but have retained the position under the label of an old "parade rest" position which has been eliminated. Since the command "at ease" requires silence, it is frequently used off the drill field at indoor assemblies and in squadrooms, shouted loudly to check conversation, and to compel attention. From this formal use, it gets into informal speech—sometimes as a synonym for the vulgar "Shut up!" and sometimes jokingly as if to say "Don't jump up and stand at attention. Pay no attention to me!"

AUTOMATIC. Originally the army name for the automatic pistol which fires rapidly without recocking, being an autoloading weapon which, so long as the cartridges in the magazine hold out, requires only that the firer successively press the trigger. Since the introduction of the Browning automatic rifle in 1917, however, this name has come to be used to describe the automatic rifle, and the old "automatic" is called simply the "pistol." But with the automatic rifle, there is a distinction to be made, between "automatic" and "semi-automatic." On the rifle, there is a lever. Turn that lever to "automatic" and the gun will continue to shoot in rapid succession as long as the trigger is held down. Just like a machine gun. But this kind of shooting is wild and ineffective in any weapon held in the arms against the shoulder. So it is the practice to limit its use to the "semi-automatic" setting of the lever, by which you have to release the trigger each time before you can press it again for the next shot. This is the principle of operation of the new Garand rifle, which may be said to reload automatically, but not to shoot automatically. Nevertheless, and also in spite of recent attempts to name the Browning the "light machine gun," many a soldier still calls it the "automatic."

A.W.O.L. In the military service it is the custom to call everything by its initials, possibly because of the set list of official abbreviations, a large percentage of which are initials, to be used on official reports and returns. A canteen with cover is abbreviated to "canteen w/ cover" and if the cover is missing, it is "canteen without cover" abbreviated to "canteen w/o cover." And so, if a soldier is away from his post, he may be "absent with leave" (A.W.L.) or "absent without leave" (A.W.O.L.), and paper records of such absence always use the abbreviating initials. Naturally, soldier talk uses them too, unless it plays with the words and says: "A.W.O.Loose" as it frequently does.

BAD TIME. Days spent in the guardhouse serving sentence or sick in hospital as a result of a soldier's own misconduct do not count toward completion of an enlistment. This "bad time" must be made up by additional days of service.

B-BOARD. A board (committee, to you) of officers, appointed to determine the qualifications of another officer and his fitness for remaining in the service. When an officer has had several unfavorable efficiency reports, he is automatically and tentatively put in "Class B" and is ordered to appear before a board of officers which decides if he shall remain in the service, or be eased into civil life with a mere modicum of his pay. They talk of going up before a B-Board, and of being B-Boarded. The process is not always fatal by a large margin, for there are many who have "beat the board." Although officially the B-Board has to do only with the separation of officers from the service, the term is also widely used regarding the "Section Eight" boards, which sit to discharge undesirable enlisted men.

BAND. Whatever a band may be on the "outside," the military band consists of horns, trumpets, cornets, trombones, drums, and fifes—all the really vigorous noisemakers. No violins. It plays marching music at parades and guard mounts, squawks tormentingly behind closed doors at its morning practices, thuds the tremendous beats of funeral notes when soldiers are laid to their final rest, and offers its members as temporary medical first-aid men during active campaigns. Occasionally, it sits around on chairs and gives periodical concerts with special musical "arrangements" like any serious civilian orchestra.

It is now being found that large numbers of troops also want classical music—which, incidentally, military bands have been accustomed to play for many years. The soldier, however, who likes the tempo of straight march-time, prefers for his regiment "a real honest-to-goodness brass band." The value of a band to foot-weary troops—and it is customary for a band to meet troops returning from a hike, and "play them into camp"—may be illustrated by an anecdote told by the leader of an army band of the "Old Army" in the Philippines:

"In 1899, at Iloilo, a battalion of the 19th Infantry marched into barracks. They had been over the mountains for fifty days chasing the scattered insurgents. Some had no shoes and their feet were swollen and bruised. Some were sick with fever; all had complaints of one kind or another. Their clothes hung in rags; they had not seen a barber or a razor all the time they were away. . . . I struck up 'Dixie' and from that 'Yankee Doodle,' 'Whistling Rufus,' etc. Each change of melody was greeted with a yell that could have been heard a mile away. Men who had limped in sore of foot and sick all over were dancing, hugging each other, and yelling like fiends."

At some military formations, when the band cannot appear, it is replaced by what is called "field music"—but we shall come to that later. A military band is a formal organization, not just a collection of odd instruments. Its birth therefore came with the rise of modern professional armies. They were first created as part of an armed force in Germany in the eighteenth century, and imitated in France. The French Guard regiments had official allotments of brass instruments in 1764. Between 1785 and 1788 (the date is uncertain) bands were formally established in the French infantry regiments of the line.

BARRACKS. "Barracks" comes to us from an old word in the French. It once meant "temporary hut or shelter." As the dictionary indicates, Edward Gibbon used it thus: "He lodged in a miserable hut or barrack, composed of dry branches and thatched with straw." Before the days of standing armies, when troops were raised temporarily and for single campaigns only, all barracks were of course temporary. The same word has been carried along and applied to buildings built of wood or stone or brick, permanent barracks for a permanent standing army. And this meaning is fairly modern, for you must recall that—except for harbor fortress troops—the British preferred not to build quarters for their troops but to quarter them in town in the homes of the people, that they might be friendly with them. Troops sheltering themselves in rows or separate little barracks, naturally came by mere habit of speech to give the word a plural form. So we always say we are "going to barracks" even now when one huge building houses a whole company or even a whole battalion of men, and even when we use the word as an adjective and speak specifically of "a big barracks building" as we sometimes do. Indeed it might now actually be correct to say that it is a word by itself, and is spelled "barracks" with the final "s" whether you are speaking of one building or two.

BARRACKS BAG. In the navy they call it a "duffle bag" in which the sailor stows away his clothing and possessions. In the army it is a denim bag in which articles of clothing or equipment may be kept, which he cannot carry in his pack or haversack. It is also convenient to keep soiled clothes awaiting a trip to the laundry. It is the soldier's trunk or valise, and troops changing station without their army trunks or "footlockers" pack in these bags all their individual belongings. In barracks, the "barracks bag" is fastened to the iron bed end, with its slip cords neatly wound and twisted around the iron cross bar.

BARRACKS 13. The "bad luck" name for the soldier's guardhouse, which is also called "Company Q" and "Battery Q."

BARRAGE. A French word meaning a "curtain" which was accepted into British and American army speech during the World War and applied to a line of bursting shells, a "curtain of fire" put down for protective purposes. For a standing barrage, the artillery fire maintains a constant danger zone across the threatened portion of a front line. A box barrage encloses a definite area as in a box, so as to keep the enemy away from it, often in order that our own troops may conduct a raid into that area and take and bring back prisoners without molestation from enemy reserves. A creeping barrage is a line of exploding shells which moves ahead of advancing troops making an attack, who follow the barrage right up and into the enemy position, while the shells keep hostile heads and machine guns down. In these more modern, recent days, we hear of an aerial barrage, maintained by antiaircraft fire to keep enemy planes away from vital objectives, and in England even of a balloon barrage, consisting of a line of anchored balloons, with wires trailing down, slight and hard to see, to snare hostile planes by tangling their propellers. The word "barrage" is frequently badly used by people outside of the service, radio talkers and newspaper writers, as synonymous with "bombardment,"—inaccurately and without regard to its curtain character. It also gets into slang, in the service, and accurately, to describe a flood of conversation purposely put up to deceive and divert. It should be added that this phrase has been supplanted officially in recent years by "defensive concentrations," although still employed in a collective sense in familiar speech.

BATTALION. A French word derived from earlier similar words signifying battle. It is therefore a group of men ready for battle. A battalion is "a little army" strictly speaking—a small group of companies commanded by a major or by a lieutenant colonel. In the United States Army, the battalion is the tactical fighting unit. In the Infantry it consists of three rifle companies and one heavy weapons company—this last used to be just a machine gun company, but now it has some mortars with it too. During the World War it consisted of four rifle companies; there was only one machine gun company to a regiment.

BATTERY. Out of the verb to batter came this word to describe the guns which would do the battering. Or it might be the place where the guns are put, from which they do the battering. The word in English is as old as 1555, adopted from the French like so many military terms. Marlowe spoke of bringing the ordnance into the "battery." In the American Army we use the word in both senses. Thus we have batteries in the Field Artillery, mobile administrative and fighting units of four guns; and we have them in the Coast Artillery at the harbor defenses with actual physical grouping in permanent emplacements which have formal names like: "Battery Parrott." The guns of a battery are rarely separated from one another. They are usually handled as a unit, and fired at a single target. Nowadays with complicated calculations for shooting at long ranges, all guns in a battery are alike. It was not always so. Note Wellington's remark (1803): "You will have a breaching battery of two 18-pounders and one 12-pounder." Never make the mistake of saying a Field Artillery captain commands a company. It is about the same size unit as a company, but its official name is "battery."

BAYONET. This word has been in our language since the days before Marlborough. It came over from the French "baionette," said to have been used to describe the stabbing piece of steel fixed in the muzzle of the soldier's rifle, because it was originally made or originally used in the French city of Bayonne. But this story may not be true. There was an old French word "bayon"—meaning "arrow" or "shaft of a cross bow"—and it is possible that the meaning was transferred from that to the piercing knife. It seems to have meant a knife and nothing more, as it was used in 1672; although if a soldier tries to use one for a knife now, he will find it pretty much too big. In 1704, we find the *London*

[21]

Gazette saying: "Our Grenadiers after two or three vollies put their bayonets in the muzzles of their Pieces." That was the way they were first attached, and the modern method of attachment is rather recent after all. So that a man would not have his rifle bore blocked and could shoot after he had affixed his bayonet, it was arranged that the bayonet could be adjusted to a knob or slot or lug on the outside of the rifle barrel. The bayonet is called "cold steel" and the French call it: "l'arme blanche" or "the white weapon." Until the invention of the bayonet, in the form of a knife inserted into, or clipped onto, the muzzle of a rifle, all infantry was composed of musketeers and pikemen, the first to shoot the projectiles and the second to rush forward for shock action. In deed the pikemen were grouped among the musketeers to protect the musketmen during the process of reloading. It was not until the beginning of the eighteenth century that the bayonet really came into its own, and the one soldier assumed the rôles of both. In 1702 in the British Army, the pike was definitely abolished and the bayoneted musket was adopted.

B.C. This is official in the field artillery for the "battery commander" and has recently become common in speech. Among the gunners, he is not "the old man" or "the captain" but is "the battery commander" or "the B.C."

BELLY ROBBER. In the old army, the company mess sergeant was everywhere and anywhere called the "belly robber" because he was always blamed for the fact that there was not enough food for one hundred men with five hundred appetites. When a soldier on kitchen police is peeling potatoes or skinning onions, he is always sure that one hundred soldiers eat more than a whole city full of civilians. The mess sergeant never starves them to death; indeed, he feeds them very well, but the soldier-given nicknames are never full of praise. And this, of course, leads us to a standing army joke:

"It looks like rain, doesn't it."

"Yeah! But the mess sergeant insists it's coffee."

Although any accuracy that "belly robber" may have had has died, the name remains.

BELT. "Pull your belt"—"Yank your belt"—"Jerk his belt." That is what they say when a man has been relieved from guard duty for misconduct. They say it because that is what happens. Members of the guard are for twenty-four hours always under arms, and always wear their belts; so that the slang phrase describes the action absolutely. The phrase also has some connection with the old army tradition that when the officer of the day chanced upon a sentinel asleep on post, he should remove the man's hat, or his rifle, or his belt, in order to have tangible proof that the man was not on the alert. The officer in such cases always brought the article removed back to the guardhouse and then sent the corporal of the guard out with another soldier, to awaken the sleeping sentinel, to relieve him of his duties, and to bring him in. The practice died out. The phrase remains.

BIBLE. Prior to the World War, all formal Army Regulations were printed in a single octavo blue bound book, which because it was the authority for and guide of action was given the slang name of "The Bible." During the 1920's, this single book was replaced by a series of loose leaf volumes of Army Regulations. To the many as to the one, the old nickname was similarly applied, and has continued in use even among the new men of the new inducted army.

[23]

BILLET. This came into English from the French *billet*. When, in less modern days, troops marched into a town and the mayor was told his people would have to house the soldiers in their barns or homes, each soldier was given a ticket which indicated where he was to go to sleep. This was the billeting or quartering of troops on the people (although sometimes of course among their cattle), which was prohibited in the Bill of Rights to the Constitution of the United States. During the World War in France—and perhaps in all future wars it will be so too—aerial observation compelled troops to get under existing shelter in buildings, else enemy planes would drop bombs on the troops or would learn from concentrations just what the generals were planning. So the practice of billeting, long disused, was resumed. Long rows of tents were too conspicuous and too easily counted. The place where a soldier slept was his "billet"— transferring the name from the ticket to the house. Arthur Guy Empey said, of his late adventure in France, that a billet is "sometimes a regular house, but generally a stable where Tommy sleeps when behind the lines. It is generally located near a manure pile." Sir H. Jenkyns, in the British Manual of Military Law, makes an interesting point. In stating that the word is a diminutive of bill (note or ticket), he points out that the military use of "billeting" for the quartering of troops, originated in England and was adopted from the English by the French. Shakespeare used it in this sense in *Othello*, Act II, Scene 3.

BIVOUAC. From the German "Beiwacht" through the French. A bivouac originally meant a night guard performed by the entire force, all on the watch. However, its meaning has changed to refer to remaining in the open all night without tent covering. In 1809, Sir John Moore wrote to Lord Castlereagh: "In two forced marches, bivouacing for six or eight hours in the rain, I reached Betanzos." So, when troops are said to bivouac on the battlefield, they spend the night in the open in their positions. Circumstances have operated, recently, to change this in our army. The experiences of the World War, when everyone was billeted under shelter, in towns or in special huts, or in dugouts, and the fact that each soldier carried in his pack the half of a small tent— these combined in practice to result in relatively little actual sleeping out under the stars—or rain!—without shelter. Consequently, in the last twenty years, there has grown up a tendency to use the term "bivouac" to describe a shelter tent camp. That is far from the strict historical meaning; but we know that the meanings of words do change with the ages. A recent dictionary of the American service shows in defining bivouac how far the meaning has broadened out, saying it is a situation "where personnel rests on the ground under shelter tents, improvised shelter, or none."

BLIND. A "blind" is a court-martial sentence which includes forfeiture of pay but does not include confinement at hard labor. A "three months blind" is thus simply a fine in the amount of part of a soldier's pay (usually two-thirds) for three months. If confinement accompanies a sentence to loss of pay, it is not a "blind" at all; it is rather a sentence of "three and two-thirds" which we shall talk about later along in the alphabet.

BLOUSE. Out in the cold and cruel world only a child or a woman wears a "blouse"—and they give the name to something that is also called a "shirtwaist." But in the army we have a blouse of our own, a military blouse. Officially it is called "one coat, service, wool, o.d.," or simply "one coat, wool, o.d." This is the "blouse" and with brass insignia and leather belt we try to cover as much of it as possible and to relieve its brown monotony with yellow glitter until we begin to look like an altar at high mass. In the old days it had a high straight collar that hooked tightly in front as if it were trying to choke each soldier by the neck, which was the despair of the hiking soldier and the delight of the ramrod staff officer. It is a coat, that is all it is. It used to be called a "tunic" and you will find that tight-buttoned article described in all the dictionaries. Yet, in spite of this, and in spite of its official designation as a "coat," you will find the American Army universally and persistently calling it a "blouse."

BLUES. The blue uniform of the army, as distinct from the khaki first introduced in Cuba and the Philippines, and the olive drab wool worn for a decade and more after 1917. It used to be standard in the service. That is why the Union Army in the Civil War was called "the boys in blue." That is why the graduating cadet at West Point sadly and gladly sang:

> We'll bid farewell to "Kaydet Gray,"
> And don the Army Blue.

But the World War put all army blue uniforms into mothballs while the troops went on campaign. They stayed there for more than ten years. Then attempts were made to revive the blues, to dress up the army, to give them a dress uniform different from their drab "working clothes." It was prescribed for aides and junior aides at the White House. It was permitted to be purchased by officers and by enlisted men, at personal expense, for wear when off duty. A few recruiting parties were dressed all up in blue. Finally after almost twenty years of edging at it, a regulation was issued which prescribed its use by officers in the evening on all army posts. The tailors did a big business. But, almost before these new uniforms had had the slightest chance to get shiny, with the expansion of the army of 1939 and 1940, blues were wiped off the regulation requirements. They are now required to be worn only at official occasions when army people visit the White House. There are a lot of them around, just the same, and a lot of them still being worn on social occasions, because the soldier does not like to "dress up" for a party in his olive drab "working clothes."

BLUE TICKET. A discharge from the army, formerly printed on blue paper, that is given in lieu of an honorable discharge (on white paper) when an enlisted man is separated from the service by reason of physical disability, sentence to imprisonment by a civil court, fraudulent enlistment, or undesirable traits of character. This form of discharge is an almost conclusive bar to re-enlistment. It is most commonly given when it is plain that a soldier fails to fit the army ways. He does not commit any offense serious enough to deserve a dishonorable discharge, but rather a succession of minor offenses which show unadaptability to military service and discipline. The procedure which is followed is described in Section VIII of Army Regulations 615-360, and is often called getting a "Section Eight." This is in the new regulations. In the old ones it was set forth in paragraph "148½," and so named.

BOBTAIL. A "bobtail" is a discharge from the army without an honorable character. In the old days, when a man had not behaved himself well while in the service, he was given a discharge on the same printed form as good soldiers with good and faithful service. However, it was then customary to write the "Character" of the soldier and of his service in a space provided along the bottom of the printed form. In such cases, he was not given any "Character" at all, and the bottom of the printed form was simply cut off with a pair of scissors. Hence the word "bobtail." This is a sample of a bit of soldier slang that depends upon an administrative fact. When the fact changed, the slang had to change too. In more recent times, when the "Blue Ticket" form was printed, and the old custom of scissoring off the bottom was discontinued, a new meaning arose. The "bobtail" now has come to mean a dishonorable discharge ("D.D.") which cuts a man's service off short of the three-year term of his enlistment.

BOILERMAKERS. The service's affectionate descriptive term for the average army band. Whether the music is good or bad, they are "boilermakers" just the same.

BOLO SQUAD. Those soldiers who at first show lack of aptitude at rifle marksmanship are put in a special group for extra training and unusually close supervision during in- struction practice on the range. This, for decades, has in the army been called the "bolo" squad. It is believed that this bit of slang came from the idea that the poorest marksmen should not be trusted with rifles at all, but rather should simply be armed with cold steel. That of course recalls the story of the recruit who missed the target four times in succession, and finally his sergeant gave him the command: "Fix Bayonet! Charge!" And also, over in the Philippines, in the days of the Empire, the soldiers had rifles and knew how to use them well, but the Filipinos were short on rifles, only their best shots had them; the others swung bolos. They were "bolomen" and so a soldier who was not a good shot with a rifle was dubbed as no better than a mere native bolo- man. He was a bolo, to be brief. The "bolo" itself, in addi- tion to being a long native knife common in the Philippines and in Panama, has been slightly altered into a straight- edged Engineer Bolo, and into a shorter more squat Medical

Bolo, both of which are articles of issue in those branches of the army.

BOLT. A part of the rifle which locks the chamber against the backfire of the exploding cartridge. Before the new rifles came in, the army carried the Springfield of 1903, and the Krag-Jorgensen of earlier days. On them, you raised and lowered a bolt handle to lock or unlock the chamber, and you pulled the bolt back by hand to extract the fired cartridge, and shoved it forward by hand to put the new cartridge in firing position. As a preliminary training for firing on the rifle range with the Springfield, there was a "bolt manipulation" exercise, in which the bolt was worked back and forth rapidly, without any ammunition, and of course without any shooting. Consequently, "working the bolt" meant making great efforts without shooting, without results, without even meaning to shoot. In organizations armed with the rifle it was common for men to speak of a soldier working his bolt, when he was just "shooting off his mouth"—without results, and usually without meaning anything much by it. He was just going through mouth exercise. Weapons may change, but you still hear of a soldier "working his bolt." In the recently motorized forces, the phrase is beginning to be replaced in the same sense by "slipping the clutch."

BOMB. The early word for "shell" fired by artillery when all early cannon were said to shoot "bombs," i.e., hollow metal projectiles filled with explosive. In 1684 at the siege of Vienna, "the enemy played on us with their cannon and bombs." John Evelyn in 1687 said: "I saw a trial of those devilish, murdering, mischief-doing engines called bombs, shot out of the mortar-piece on Blackheath." During the World War the hand grenade was sometimes called a bomb, and the grenade thrower was more likely in the British Army to have been called a "bomber" than a "grenadier." This may have seemed strange but it was due to one of those merry-go-rounds of history as we shall see when we come to the word "grenadier." Since that war, however, there has developed a new and important use of the term bomb, one strictly in accord with its original definition. Airplanes were developed specially to drop "bombs" on enemy installations and troops. These were called "bombardment" planes or "bombers." So, with the term dropping out of the artillery where it originated, and being forced upon the trench fighters of the doughboys, it was taken up and adopted by the aviators, so that nowadays one scarcely hears or sees it used except in connection with aerial fleets. It is a short word, its very sound descriptive, almost identical in English, French, German, Italian, Latin, and ancient Greek.

BOMBARDMENT. Although the word "bomb" has been dropped by the artillery as a word describing their shells, it has been retained by them as part of the word "bombardment" to describe heavy fire concentrated upon a locality. The weeks of artillery shelling which preceded some attacks during the World War were called "preliminary bombardments" as also were some of the shorter ones staged late in 1918. An Old French noun *bombarde* (cannon) developed a verb *bombarder* and "bombard" in English. The suffix "-ment" was added to indicate the act of bombarding and it became a noun again. Thus words go around the circle of grammar. The verb, it is noted, is a transitive verb, and must, by the ancient rules of rhetoric, have an object. A "barrage" is fired whether or not there is an enemy where it will fall; its purpose, as we have seen, is to make a curtain or barrier of fire. A "bombardment" on the other hand is aimed at positive destruction of definite places.

You bombard a city, or bombard a fort. Technically, the artillerists do not use the term very much; they use it only in informal speech; and it appears more in novels and newspaper accounts than it does in the written regulations of the army. The official language gets into such details as "preparatory fire" and "barrage" and "harassing fire" and "interdiction fire" and "concentrations" and "destructive fire" and "fire for effect" and clings to this military jargon, eschewing the forceful and effective "bomb" and "bombardment."

BOWLEGS. This is a somewhat derisive term for a cavalryman, and like all true slang is descriptive and figurative in origin. Because constant riding of horses is thought to bow his legs, the cavalryman has for years been called Bowlegs. Just because many have taken to armored cars and fighting tanks in the mechanized brigades and armored regiments, do not imagine that all horse cavalry has been eliminated from the army. In certain kinds of country unsuitable for motored vehicles, it cannot be replaced. Some remains. So long as it remains, the cavalryman will continue to be called "bowlegs."

BRASS HATS. So far as we can find, this came into the American Army by straightforward transfer from the British during the World War. British staff officers wear army caps with a great deal of gold embroidery. They are therefore "brass hats" and were so called by Tommy Atkins in France. Our staff officers wore no such decorations as such, but they got the name "brass hats" just the same, and have kept it ever since. There is an extension of this use, however. The American tends to speak of the commanders, the generals as well as their aides and staffs, as "brass hats." The term is not exactly coveted, for behind its adoption is the ancient jealousy between line and staff. Those who wear the "brass hats," it seems to say, are all dressed up while the poor boys with the troops do the dirty work.

BREAD SERGEANT. A satirical title given the private soldier detailed as dining room orderly, one of whose main duties is to slice and serve the bread.

BREAK. An interval between the completion of one enlistment in the army and the beginning of another. For example, a "three-day-break" would be three days between a discharge and re-enlistment, three days of freedom as a civilian, until the lure of the service brings the soldier back earlier than he had thought when he drew his final pay. Thus, also, any interval is a break; and short rest periods between instruction hours, on the drill ground or in service schools, are commonly initiated by the officer in charge looking at his watch and saying: "There will be a five-minute break." We never use the word "recess."

BREECH. In the plural, this might make up something you wear on your legs and your seat. In the singular, however, it is part of a gun. It is the technical name given in the army to the rear end of the bore of a rifle, closed, when you get ready to shoot, by the bolt or, in the guns of the artillery, by what is properly called a "breech block."

BRIGADE. Originally meaning a group or company, coming from the French "brigade" from the Italian "brigata" (a crew of fighting good fellows), this has become a technical army term. A brigade consists of two regiments or three, and is commanded by a brigadier general. In England they call him simply a "brigadier." The number of regiments varies with time and place and with special organizations. In fact, in the new 'triangular" division of the United States Army, the brigades have been eliminated entirely, although they still exist in the cavalry and the corps artillery.

BUBBLE DANCING. This is the slang way of a soldier speaking of washing dishes in the company kitchen when he is on "cook's police." That is, unless he calls it "pearl diving." Sometimes in place of being called a pearl diver or a bubble dancer, he is dubbed, according to the army's habit of uncomplimentary cognomens, a China Clipper. This last is, of course, very recent, but is gaining ground rapidly.

BUCK. A buck is a private soldier as distinguished from a noncommissioned officer. The word is mostly used in the phrases "buck private" or "big buck." It really means one who wears no chevrons, so even the first-class private with his single-striped chevron is not really a "buck."

BUCK, PASSING THE. Passing the buck is shifting responsibility or blame. If something is to be done or decided, it is passed on to someone else of junior grade. If there is blame to be taken for anything done improperly, it likewise is passed on down the line. If a subordinate is not equal to making a decision, he asks higher authority and passes the problem along, thus passing the buck up. The more common use and meaning, though, has to do with passing the buck down. George Colman the Younger described it in *The Iron Chest:* "The rubs I take from him, who is above me, I hand on to you who are below me. 'Tis the way of office: where every miserable devil domineers it over the next miserable devil that is below him." The quotation is out of England and a hundred and fifty years old. Mark Twain used the phrase in 1872, but in another sense, in a poker game, and not in the army sense. Wallace Irwin had a politician "pass the buck" of responsibility in 1912, and it is only right to say that the army has used the phrase so constantly in the twentieth century that civilians even speak of it as "the old army game." An example of the way it works in the ranks is told in William Hazlett Upson's *Me and Henry and the Artillery* as follows:

"Here came the captain shouting and hollering to the looeys: 'Break camp! March order! We move out in one-half hour!' So then the looeys started hollering to the sergeants. And the sergeants started hollering to the corporals. And finally the corporals started hollering to the privates, which was us, and they kept hollering and we did the work. Privates ain't got nothing to holler at, unless you're a driver, which we wasn't, in which case you can pass it along by hollering at the horses."

Then the practice was expressed in an army rhyme:

> The captain told the lieutenant
>   To polish up the floor;
> The lieutenant told the sergeant,
>   And, Gee! but he got sore.
> The sergeant told the corporal
>   Who got mad as he could be;
> I've just talked to the corporal,
>   So I guess it's up to me.

BUCKING FOR ORDERLY. It is the custom in the army at guard mount each day to select from the detail for the guard the best dressed and best appearing soldier, who becomes orderly for the commanding officer for the next twenty-four hours. As such, he sits during daytime at the door of the commanding officer, and carries a few messages, and then at night—when other members of the guard are walking through the darkness or cold of the night—he goes back to his bunk and gets a normal night of sleep. As a special reward, he also usually gets a twenty-four hour "pass" or release from all duties. So, when a soldier—mindful of these possible perquisites—makes special efforts to improve his appearance, to see to every minor detail of polish of brass, press of uniform, and trim of hair, he is said to be "bucking for orderly." In a great many instances organization pride at having one of its members selected for orderly in competition with men of other companies is so great that an entire outfit will assist the man most likely to win. Sometimes, tradition says, they will even carry him bodily to the place where the guard is going to form, so that he will not get any dust on his shoes. Incidentally, a man too regularly misbehaving is said to be "Bucking for Section Eight."

BUCK SLIP. When an official piece of correspondence comes to the adjutant's office, he passes it on to the particular staff officer charged with the matter which it covers. He does this usually informally and briefly by hitching onto it a small slip of paper, printed in advance, on which by mere check marks he can indicate to whom the document should go and what must be done about it—for draft of a reply, for investigation, for information only, and so on. Because of the army slang term for "passing the buck" and because the adjutant is just passing the paper along, this form slip is pretty generally called a "buck slip."

BUDDY. The American soldier is not a sentimentalist. The word "friend" is too formal. "Comrade" is too gushing. In the A.E.F., he adopted the term "buddy" for his friends. It was widely used, to speak of them, to hail them. It was used even to strangers: "Hey, buddy, what outfit?" With the concurrence of the American Legion, the word has come to be understood to mean "a soldier." "A buddy" is a soldier. "My buddy" is my friend in the service with me. Frederick Palmer says that the word is confined to privates, and that he never heard a general refer to himself as a buddy. He didn't say so, but he didn't hear any privates referring to a general as a buddy either. The general use of the word arose in the A.E.F. of the last war, but it has been known in this same sense in America before that—in West Virginia mining areas where the two workers who cut at coal up a heading together, and worked in pairs regularly, spoke of one another as "buddy." O. Henry put it in the mouths of cattlemen in 1904. There is another use of the word in America, as a family nickname for "brother," and Kenneth Roberts tells me that General Francis Marion of the Revolution used it to describe his half-brother and body servant. This does not, however, on account of its family connotations, appear to have been the origin of the military use of the word.

BUGS. Our modern mechanical army is using this word a lot. (No! It has nothing to do with "cooties.") It was brought into the service by the motor hounds, the men of the motor transport schools and the tank school. It is a pure borrowing from civil life. A "bug" is something in the nature of a defect of design or of adjustment in a machine. If a shaft is not strong enough and is likely to break, if a bearing is not correctly designed to carry on without failing, if a motor tends to overheat under its normal load, if the gasoline supply line runs in a dangerous or vulnerable place— these are all "bugs" and you have to get rid of them before you will have a machine that will stand the hard knocks and raw usage of military service. You cannot get rid of these "bugs" the way you can get rid of "type-lice" either; you have to change your design and build over again, part way at least. The maximum number of "bugs" in the United States Army was in the days when we were, between 1925 and 1935, experimenting with a series of "pilot model" high-speed tanks. I suppose the aviators have "bugs" too, and the Signal Corps, but I happen never to have heard them use the term. The rest of the army uses it everywhere and all the time, and applies it not only to mechanical defects but also to errors in paper plans, schedules, or orders, and to imperfections of any sort.

BULL. Short in soldier speech for "bull's eye"—indicating that the rifleman has struck the black center of the target. The soldier of course also uses the word in the same slang and uncomplimentary meaning that ordinary civilians do, as "throwing the bull," but this naturally has no relation to target range or marksmanship. The same word thus has two totally different uses: one for ineffective chattering and the other for effective shooting. In the latter sense it is one of the happiest words for any soldier to hear.

BULL PEN. Originally a fenced enclosure for prisoners. Parson Weems used it in this sense in 1809 in a life of Francis Marion. The army still uses it, even as a synonym for a solidly built guardhouse.

BULL RING. The closed and fenced corral on a mounted post where horses are exercised and training is given to horses and men, as distinguished from training on the drill ground or across country. So far as we know, cattle never see it. During the World War, according to S. Eddy's *With Our Soldiers in France*, the name was used to describe a training area in rear of the front lines, where divisions were given their final licks of polish before being moved up to participate in an attack. This use frittered along, in and out of popularity for two decades. It has been now revived at the recent training camps, to designate the separate training areas for recruits, and also for the official Replacement Training Centers where drafted men spend thirteen weeks before being assigned to their regiments.

BUNK FATIGUE. One soldier has described this as "Wednesday afternoon, Saturday afternoon, and Sunday on a bunk," which is what the soldier calls his bed or cot just as do many civilian cowboys, woodsmen, and construction workers. There were "bunks" of wood or straw for troops so spoken of as early as 1758. The "fatigue" part of the name is ironic, for "fatigue" is practically the official army term for hard work, as we shall see later. "Bunk fatigue" is the only kind that soldiers really like. They take the others as they come, but never with relish. As D. F. Rowse has said, bunk fatigue is always taken with pleasure, but it must never be confused with ordinary nightly sleeping. The soldier uses it only to describe time spent on a bunk in the daytime.

BUNK FLYING. Down in Texas at the flying schools, they use aviator language when they concoct their familiar slang. When the infantryman will speak of "working your bolt" in idle speech, the young flyer will speak of "bunk flying" because perhaps it is done on the bunk, or is perhaps just "bunk" in the general American slang sense of the term. The aviator will say that it applies to any talk of aviation in barracks, instead of performing aviation outside.

BUST. Reduction in grade. When a noncommissioned officer is reduced from his sergeancy or corporalcy, or a first-class private from his grade to that of private, he is said to be "busted." If he receives a sentence of court-martial with confinement and a reduction in grade, he is said to have received, for example: "Three months and a bust."

BUTCHER. The soldier gives nicknames to everyone he can, and the nicknames are never complimentary—no offense meant either. So he calls the company barber, a soldier who

shaves faces and clips hair on off hours as a standard con-
cessionaire of the company—calls him the "butcher." Then,
when you get around the hospital, you will find that this
title, in still less dignified form, is commonly assigned to the
chief surgeon, as just "Butch."

BUTT. A remnant, or small piece, as the butt of a
cigarette. It is often used in the army, however, to mean the
entire cigarette, as "Lemme a butt" meaning "Let me have
a cigarette." The basic idea, though, in the army as on the
outside, is that the "butt" is a small piece or remnant. From
this comes a truly military slang application of the word. It
is used by soldiers to speak of part of the time still to be
served on their enlistment. "He has three months and a butt
to do" means that the soldier has three months and a fraction to serve before his time comes for discharge. Likewise,
"a year and a butt" means a year and a fraction—the "butt"
being perhaps more, perhaps less, than a single month. This
word, as also "meat ball" and "roll over," naturally dropped
out of use in the drafted army of 1940 and 1941, where en-
rollment was for "twelve consecutive months" and more
by consent of Congress. It is firmly fixed in the vocabulary of
professional enlisted men, however, and will probably return
when circumstances restore the occasions for its use.

BUTTS. Note that this has the final s, and it is always
used thus. Strictly speaking, the "butts" are the embankment
set to stop the bullets at the end of the target range and to
protect the men behind them who manipulate the targets.
"You are firing into the butts" means that your shots are
low and are hitting the embankment. It is sometimes loosely
used to indicate the pits behind the embankment, and sometimes even broadened in meaning to apply to the entire
target end of the range as distinct from the firing points.

But this generalization is a common characteristic of all words, whose meanings tend to broaden as time and careless use carry them along.

BUZZARD. A discharge from the army. Indeed, even specifically the paper which testifies to that discharge, the paper that is which is called the "Army Discharge," although it is of course not the discharge itself, but only the evidence of it. If you put this into the plural "buzzards," it has a totally different meaning. It is then a familiar way of referring to the two silver eagles, insignia of rank, which adorn the shoulders of the colonel. Not that they look like buzzards at all, but simply that the soldier likes to give less dignified names to anything that meets his eye, or that has a serious aspect. He lightens with his speech every phase of the life which surrounds him. This same trait, of course, is seen in the labelling of the formal and important discharge paper a "buzzard," because it has a large representation of the official American eagle emblazoned at the top of the sheet.

CAISSON. From the French word, similarly spelled, which means "large chest." It came to the American Army by way of the British and the language at large. Specifically it means the large ammunition chest, drawn on two wheels by horse artillery. There still is a little horse artillery in the army, and still a few caissons. But even if there were not, the term would endure, for it is implanted in the most generally popular army song we have. This version of the song differs somewhat from the wording sometimes encountered, but this is the final version of the copyrighted song:

> Over hill, over dale,
> We have hit the dusty trail,
> And the caissons go rolling along.
> Counter March! Right about!
> Hear those wagon soldiers shout
> While the caissons go rolling along.

[40]

The word is characteristic of the Field Artillery, and even ammunition trucks will probably not stamp it out. The foot soldier, proud of his marching abilities, uses the term as a sneering nickname for the artillerymen, calling them "caisson riders."

In a mechanized field artillery there are no caissons, but the old song continues. Its author, General Gruber, is immortalized at historic Fort Leavenworth by a caisson set in stone on a commanding hilltop.

CALLS. A tune blown on a bugle is labelled a "call." Since the bugle blows to bring soldiers to assemblies for duty, these assemblies themselves are by metaphor spoken of as "calls." Although bugles have been known in armies for long ages, it was not until the Civil War that they were fully standardized in the American Army. Prior to that time regimental commanders often used the drum instead of the bugle, indeed it was generally used for "Taptoe" or "Tattoo" as we shall see. Certain special "calls" do not refer to ordinary formations. One called "Overcoats" is a warning that men should wear overcoats for the particular formation for which the call is being sounded. With regiments collected at a single station, it has frequently been the practice for each regiment to have its buglers sound a bar, or a few additional notes, characteristic for that regiment, to prevent other, adjacent regiments from mistaking the call for them. These were most frequently used with "Officers' Call." They were informal and were only irregularly adopted until 1940, when the War Department specifically authorized them.

The music of some of the better known calls is given in this book under the following headings:

Adjutant's Call	Mess	Reveille
Assembly	Officers Call	Sick Call
Colors	Quarters	Stable Call
Fatigue	Recall	Taps
First Call		

CAMPAIGN HAT. The stiff, flat-brimmed, peaked hat typical of the army almost since the Spanish-American War, is officially named the "hat, service," but from its early adoption for field use in tropical countries it has been familiarly called the "campaign hat" to distinguish it further from the leather visored cap which many have called the "barracks cap."

CANNED WILLIE. This name for corned beef hash has a long and honorable history in the speech of the American Army. The A.E.F. took it to France with them and brought it back, and then distributed it over the country in the ranks of the American Legion. If it can be said to be general slang, it was specially military in origin and remains military. It was immediately taken up by the drafted men of 1940.

CANNON. This, in origin classical, then French, comes to us from "canna" meaning a reed, pipe, or tube, and written "canon" and "cannon" meaning a great tube. In 1525 T. Magnus mentioned five "gret gonnes of brasse called cannons." And we of today all remember Tennyson's "Charge of the Light Brigade." In recent years the word has not been used much by military people. All modern "cannon" are called "guns" or "howitzers" or "mortars" officially, and "pieces" informally. The modern "gun" is rifled, with a spiral twist in the bore to make the projectile spin, and "cannon" is employed only to refer to the antiquated smoothbore gun of Civil War days and before, which threw "cannonballs."

CANNONEER. A soldier who serves with an artillery gun or cannon. It is old. A law of Queen Elizabeth of England mentions "gunners, commonly called Canoneers."

It is still an official term, even though the name of the weapon has been changed from "cannon" to "gun."

CANTEEN. A French *cantine* is a cellar or a small shop, and in Spanish countries a *cantina* is a little country store. In vulgar French the *cantine* is also a small vessel, and the word thus got into the army and the English language to represent the metal flask in which each soldier on the march carries his own supply of drinking water. A song which was once very popular with the Grand Army of the Republic, written by General Charles G. Halpine, depicted comradeship with the repeated line:

> There are bonds of all sorts in this world of ours,
>   Letters of friendship and ties of flowers,
>   And true lovers' knots I ween;
> The girl and the boy are bound by a kiss,
> But there's never a bond, old friend like this:
>   We have drunk from the same canteen!

Although *bidon* has taken the place of *cantine* in the French Army, the British and Americans cling to and have made official the word canteen. There is, however, another "canteen" in the army. For many years the British Army has maintained government shops for the sale of beer, other liquors, and minor supplies for the personal needs of soldiers. In America, dependence at frontier posts had been upon occasional merchants, or—later—upon officially sponsored "post traders." In 1880, however, General Morrow and some officers of the 21st Infantry at Vancouver Barracks established for their regiment a cooperative store and, imitating the British, called it a "canteen." It succeeded. By General Orders No. 10, of the army, February 1, 1889, rules and regulations were promulgated for the establishment of similar canteens in other posts. By February 1, 1890, there were fifty-two in successful operation. The "post trader" system was abolished. By G. O. 11, February 8, 1892, the designation "Canteen" was officially changed to "Post Exchange" but the soldier kept on calling it "The Canteen" and probably always will, except when he uses the abbreviation "P.X."

CANTEEN CHECKS. A large part of the business of a Post Exchange used to be done on credit. Officers and some enlisted men might have charge accounts. Others were periodically issued books of credit coupons; but these were never called "coupons"—always "canteen checks."

CANTONMENT. A semi-permanent camp for troops, now pronounced in the United States Army with the accent on the second syllable. In the English language at large, the word was long used to refer to the scattering of troops for billets or quarters over a large area, as when they went into winter quarters, as distinct from their concentration for campaigns. You will find it so explained in Farrow in 1883. It was also definitely used to refer to temporary barracks built for troops, as the cantonments along the Thames in Barham's *Ingoldsby Legends*. In America, where it was used in 1777, it was constantly used to refer to established army posts, to which they did not want to give the word "fort" because there was no fort there. Until 1832 it was common in this sense; Fort Leavenworth, Kansas, was first called a "cantonment." Its last appearances in army orders in the nineteenth century occurred when the cantonment at Fort McKinney was renamed "Fort Reno" in 1877, and the "Cantonment of the Uncompahgre" in Colorado was renamed "Fort Crawford" in 1886. The word practically dropped out of American Army language for twenty years at least. It was reapplied by acts of Congress in 1906 and 1907, appropriating money for "summer cantonments" and for the hire of buildings and grounds for summer maneuvers of the regulars and guard combined.

In 1917, when divisions were being organized for overseas service, many of them were quartered in leased areas officially termed "camps" although on some of these temporary wooden buildings were substituted for the usual tent camp. Their design was based on a theoretical life of five years. These wooden cities were officially called "cantonments" and the word was popularly used to distinguish that sort of place from a tent camp. As a fact, the distinction got so fixed that General Orders which had used the phrase "divisional camps" were altered to read "divisional camps and cantonments." It has been used since the World War to refer to buildings erected for summer camp use only. So, its old meaning as an act of spreading troops over a large area, perhaps into distant "cantons," has now been superseded by a meaning for a single temporarily built-up place where troops are concentrated.

CAPTAIN. From the Latin word "caput" meaning head, this word indicates the chief or head man. Head of what? Well, mere custom, extended over centuries, has settled that he shall be head of that particular military unit called a company, troop, or battery. He might have been head of a battalion, or of a regiment, but custom pinned him down to a company, and there he is. (Until he gets promoted!) In the seventeenth and eighteenth centuries there was a title captain-general frequently used. Thus Marlborough was captain-general of the Dutch allies fighting against France in the early eighteenth century. And the 1776 Constitution of the State of New Jersey specifies that the governor of the State shall "act as captain-general of all the militia." Pinning the single term "captain" down to a company has ancient authority. Muster rolls of the British in France in the 1420's use the word. Barrett in 1598 speaks of a "regiment [divided] into companies, over every company a Captaine." The captain is the highest responsible officer in full command of a unit who still has direct and constant contact with all his men. He knows all about their shoes and socks and haircuts and chow. He keeps them straight and helps them out. The captain is the leader of his company. What Boulton said in 1618 is still true; "Such as the Captaine is, such is the Souldier." Shakespeare, in *Measure for Measure*, seems to think he has privileges:

> That in the Captaine's but a chollericke word,
> Which in the Souldier is flat blasphemie.

CARABAOS. A nickname given one of the regiments of the American Army as a result of a confusing incident in the Philippine Islands during the days of insurrection. A portion of that regiment (which prefers not to be named) was alarmed and stampeded one dark night from a foundationless fear of a hostile attack on its encampment, which arose simply because sentinels fired on and were frightened by a herd of water buffaloes or carabaos. The name has become a humorous tradition in the United States Army, and an organization of officers who served in the Islands during the insurrection has taken it as part of its official designation: Military Order of the Carabao. The carabao himself is an ugly, humpback beast somewhat like a lighter and very dejected ox. He is the subject of a verse of a traditional song of the Islands:

> The carabao have no hair.
> They run around quite bare,
> For the carabao have no hair in Mindanao.

CAVALRY. Coming from the French *cavallerie* out of the Italian, this means horsemen or a military force mounted on horses. It has its origin in the Latin word *caballus* (horse) and connects up with the word "chivalry" which was a courteous set of customs in days when knights in armor were mounted horsemen. In those olden times, "a body of chivalry" was used for "a body of horsemen." The French simplified the word down to *cavalerie* before the British took it over and clipped it down, and we of course got it from them. Until 1940, we had in the American Army a brigade of cavalry called a "mechanized brigade," made up of historic cavalry regiments like the 1st Cavalry and the 13th Cavalry, but without a *cheval* in sight! This anomaly has disappeared, because its regiments and the brigade have been made into and renamed "armored regiments" and the "armored division."

CHANNELS. "Through channels" or "through military channels" means the routing of orders, requests, and reports through all intermediate commanders and headquarters. For example, a letter from an individual which is sent "through channels" passes in succession for remark or recommendation through company, regimental, division, corps and army commanders before being referred to the adjutant general. It takes time, thus, to send a letter from Fort Sam Houston to Washington, but each intervening officer has a chance to give the proposal a knock or a boost, as he sees the matter from his angle and in the light of his responsibilities. After all, the colonel is supposed to have something to say as to what is going on in his regiment and what his officers are proposing to do. Else why have a colonel? (And I wrote this when I was a captain!)

CHAPLAIN. The title "Chaplain"—the word being derived from the Latin "capellanus"—had its early origin in the "Cappa" or "cappella" of St. Martin of Tours which became Old French "chapele" and the English "chapel." It is related that St. Martin gave half of his military cloak or "cappa" to an importuning beggar at the gate of Amiens, and wrapped the remaining portion about himself as a cape or "cappa." Tradition affirms that this cape, or its counterpart, was preserved as a relic by the kings of France and taken with them as a talisman when they went to war. The tent where it was kept was guarded by special custodians called "capellani" who celebrated services there by the relic as in a church. From this sequence of events, military clergy came to be called "chaplains."

The chaplaincy in the United States Army commenced during the Revolution when George Washington issued the first call to the colors for American ministers of the gospel. They served for six-month periods, with the pay allowance of a major, but without actual rank. Many of the best known clergymen of the day took their terms with troops at the front or ministered to those in hospitals or prisons. Although the organization of the United States Army dates from September 29, 1789, it was not until the Act of March 3, 1791, that the office of chaplain was officially recognized as a part of the armed forces. In an army of 2,232 officers and men, there was to be one chaplain, to be appointed when the President deemed it in the public interest.

History tells of ten complete reorganizations of the army; in four of these Congress completely forgot to provide for chaplains! During our various wars, however, chaplains were always in evidence. In the War with Mexico, there was one chaplain provided for each regiment. In the Civil War, there was one for each regiment of volunteers, and these chaplains worked hard. In fact, one of the features of the Civil War was the succession of religious "revivals" with which souls were saved and men kept out of mischief during the dull days between active campaigns.

CHEVRON. An inverted "V" worn on the arm. From French it goes back to Latin *Capreoli*, which was applied to two inclined beams meeting like rafters. In heraldry it denotes an upside-down "V" on a shield. Now it refers to the V-shaped stripes on a soldier's uniform. There are wound chevrons, and war service chevrons, as well as chevrons to denote rank. Lord Roberts speaks of the chevrons "to mark certain periods of good conduct." In our army as early as 1782 a service chevron was adopted for wear on the left arm, one chevron for three years of service. In 1863, when the terms of some regiments were about to expire, an order was published that men who had served two years might re-enlist as "veteran volunteers" and each—in addition to other rewards—be granted the right to wear a "service chevron" to "make him proud in the presence of the humble recruit." In 1904, pending the issuing of campaign badges, the War Department authorized certain temporary "service-in-war" chevrons, to commemorate service in 1898 and 1899. During the World War, chevrons were also devised to show service in France.

Chevrons to denote rank were not originally limited to enlisted men. (Indeed, they are used to show grades of cadet officers today at West Point.) The custom of wearing the inverted "V" as a mark of leadership is said to date from medieval times, when the lord of the castle granted it to his most prominent workmen, who had perhaps helped raise the chevrons of the roof rafters, we might say. From this, in the society of the time, it was a short shift to use it to denote military leaders. It is a practice common to all armies. Wellington in 1813, for example, speaks of "serjeants' chevrons." They appear in our regulations in 1821, captains and lieutenants wearing them of gold or silver, sergeants and corporals of worsted. They used to point downwards, but since the Spanish-American war period have pointed upward. A first-class private wears a single chevron, a corporal two, a sergeant three. Staff sergeants have one, technical sergeants two, and first sergeants and master sergeants three lower loops under the triple chevron.

CHICKEN. A colonel wears on each shoulder a silver eagle, and these have fondly been termed "chickens." They have their mighty glory, but the soldier says:

> In place of being colonel
>   With a chicken on my shoulder,
> I'd rather be a private
>   With a chicken on my knee.

CHOP CHOP. This is pidgin English and comes from the Far East, and the garrison duty of American troops there. It means "quickly" and is common in the Shanghai and Canton areas, as also in the Philippines. It has had only a sparse use in the Regular Army, but the recent draftees seem to have taken it up.

CHOW. This is the army way of speaking of a meal. It is said by Mencken to have been introduced into the United States with the first arrival of Chinese in California in 1848, and is thought to be an Americanized corruption of the Chinese *chia*. The word was current on the Pacific Coast for many decades, but has pretty well died out in civil life in the country as a whole, although it has taken a fast and firm hold in the army. A "chow hound" is first in line at mess.

CITS. The army abbreviation for "citizen clothes," as distinct from uniform.

CIVVIES. This runs the preceding word a close race in popularity as an abbreviation of the words "civilian clothes."

These of course may not be worn on a military reservation by members of the service except when arriving or leaving in connection with a trip to town, and except for civilian evening clothes, or civilian golf, riding, or tennis clothes when engaged in those sports. During the World War, the wearing of civilian clothes was forbidden to all military personnel, whether on a military post or off, whether on duty or off duty. This prohibition was relaxed after the Armistice of 1918 but was reestablished in 1941.

CLOSE ORDER. This was a formation, under the old drill regulations, in which men stood in line in two ranks, and performed the old movements by squad units: squads right, right by squads, squads right about, right front into line, etc. etc. There were only four inches between men in line. It was called "close order" as contrasted to "extended order" in which each squad stood in single rank with intervals of five yards between individuals. New regulations adopted in 1938 completely wiped out the old elbow-to-elbow form of "close order drill" with its squads right, squads left, and the rest, and had soldiers form at arm's length intervals and march always in column. But it did not destroy the distinction between "close order" and "extended order"—the latter being retained to distinguish the informal movements and deployments for battlefield use. The difference between the two was lessened, but not destroyed.

COAST. Everything is abbreviated in the army, it seems, or nicknamed, and "Coast" is short for Coast Artillery, that branch which mans the harbor defenses and permanent fortifications along the seaboard. During the World War, the Coast Artillerymen did not want to sit and look out to sea all the time; so they dismounted some of their big guns from emplacements that would probably never get a chance to fire at a big ship of the German Navy, and took them to France, to be formed into brigades of heavy artillery to support the army attack in the Meuse-Argonne. In addition to handling the antiaircraft work, the Coast Artillery now is charged with the mobile big guns, railroad and tractor drawn. In spite of these recent tasks, that branch still suffers from that good rivalry and raillery between branches, which developed years ago this libelous parody on the caisson song of the Field Artillery:

> Then it's hi, hi, hee, the Coast Artillery,
>   The branch of the women and the wine.
> In my cottage by the sea, I will sit and sip my tea,
>   While the didies hang out on the line.

COFFEE COOLING. Loafing in comfort while others work, or perhaps having an agreeable job while others are out enduring the heat and the dust of hard field service. It is used to describe men on certain inside or desk jobs, as distinguished from those who have to do drill or participate in training maneuvers or campaigns. One of the regiments of the United States Army (which prefers to forget the fact) is nicknamed the "coffee coolers" by others because during the Philippine insurrection that regiment remained in Manila while other regiments were engaged in active operations against the insurrectos. During the Civil War the alternative term "coffee boiler" was more frequently used, and with a slightly different meaning. The type is described in Morris Schaff's book, *The Battle of the Wilderness*:

"A real adept skulker or coffee boiler is a most interesting specimen, and how well I remember the coolness with which he and his companion—for they go in pairs—would rise from their little fires upon being discovered, and ask innocently, 'Lieutenant, can you tell us where the Umsteenth Regiment is?' And the answer, I am sorry to say, was too often: 'Yes, right up there at the front, you damned rascal, as you well know!' Of course, they would make a show of moving, but were back at their little fires as soon as you were out of sight."

COLONEL. The colonel was so called, according to Skeat, because he led the little column at the head of the regiment. Spaulding believes the word stems from a little column or "colonello" in Italian, formed for independent missions. When permanent organization began to be made, there appeared a "colunela" or group of companies commanded by the senior captain, designated as "cabo de colunela." Then the word was corrupted in Spanish and in early English into "coronel"—possibly by some analogy with "corona" or crown, for these were royal troops. But in England the spelling with the "l" was resumed, even though the pronunciation still bears a trace of the early form: "ker-nel." Since Elizabethan days in England the modern form has become fixed. In 1598, Barrett said that the "Colonell is the commander of a regiment" and in 1814 Wellington specified: "The regiment to be commanded by a colonel and each of the battalions by a lieutenant colonel or a major."

COLONEL'S WIFE'S SILK STOCKING. This, among the aviators, is what they occasionally call the "wind stocking" flown in the breeze to indicate the force and direction of the wind.

COLORS. The origin of this word is too obvious for explanation, its use being a mere shortening of the idea of a colored flag to colors. In the army, however, "Colors" bears a status more significant than a mere nickname. There is a vital and important distinction between "Colors" and a mere flag. It is a flag which flies from the flagstaff in an army garrison. It is a flag which hangs across a street on a holiday. "Colors" are flags specially entrusted to an organization. There are national colors and regimental colors. They are carried together, and always brought out under armed guard. They rate a salute from everyone; a mere "flag" never does, not even a national flag. The national colors are of course a sort of special edition of the flag of the United States. Although there were occasions prior to 1913, they are now never dipped in salute. (The British do dip theirs; indeed they even slant their staffs on to the ground in homage to royalty.)

The regimental colors bear a conventionalized American eagle, with an emblazoned shield showing by the patterns of heraldry some device or devices symbolic of the history of the regiment, all against a background of dark blue for Infantry, red for Artillery, yellow for Cavalry. On the regimental color are embroidered the name and the motto of the unit. On the staff are fastened streamers on which are lettered out the names of historic battles for which the unit has been officially awarded battle honors. Regimental colors do dip in salute. They dip to the tune of the "Star Spangled Banner" and they salute an officer or official of higher rank than the commander of their regiment: it is as if the regimental colors themselves joined the regiment in paying honors where honors are due.

"Come and Get It!"

"To the Colors" is the name of a bugle call, which with field music is a complete official substitute for the national anthem played by a band. Since 1907 in the mounted services—that is, in the Field Artillery and the Cavalry—"colors" are somewhat smaller in size, and are borne on staffs somewhat shorter, and are called "standards."

COME AND GET IT. The sweetest words of army life. Whatever the bugler may do with "mess call," the lads in line must wait for the mess sergeant to call the company to come to a meal with the time-honored, traditional words: "Come and get it."

COMMAND. A command is an oral order. At one time its use was restricted to those commands specifically laid down in the drill regulations, but now it has a somewhat broader use. "Command" is also used to denote any military unit. Thus a company is an appropriate "command" for a captain, a battalion for a major, and so on. Also, a military post, or even a district, is said to be a "command" because it has a commanding officer. This use of the word has long been common, although unofficial, in America. In England it is official, and they speak of "The Aldershot Command" and "The Home Command." Recently it has been made official here, with the creation of the Alaska Command, the Service Command in each Corps Area, the Newfoundland Base Command, and Caribbean Command. The word is also used in another set phrase. We say: "Who is in command here?" instead of saying "Who commands here?" "Commandant" is rare in the American Army. Actually synonymous with "commanding officer" it was so employed from 1798 to 1867, at least, but now is used only in special circumstances, as for instance to describe the "commandant" of a Disciplinary Barracks, of military units at civilian colleges, of cadets at West Point, and of regular army schools.

COMPANY. A general word with a special military meaning, referring to a definite unit of infantry soldiers. In the cavalry the corresponding unit is called a "troop"—in the artillery a "battery." Until 1883 there used to be "companies" of cavalrymen in the United States Army, and indeed until 1883 officially even "companies" of Field Artillery, although the term "battery" was very common for such units during the Civil War. Until 1924, the batteries of Coast Artillery were termed companies, too. A company, in the general sense is a group of men together, men perhaps who together (com) eat bread (panis). It has always been so termed in modern armies, and constant usage has settled this word with a very general sense at bottom, into a fixed term for a unit which lives in the same barracks and eats at the same mess. It is the basic administrative unit of the army. Barrett in 1598 said: "The Campe-maister deuides his regiment into companies." Thus in each regiment we have a succession of companies designated by the letters, A, B, C, D, etc.

COMPANY PUNISHMENT. For minor misdeeds, under full authority of a legal enactment in the Articles of War, the company commander may award certain limited punishments without the interposition of a court-martial. The soldier has a right to demand trial, if he wishes. Before he can be given company punishment, he must admit the offense and state he is willing to accept company punishment. The captain can then formally award him a task of extra work around the barracks, in the kitchen or on the coal pile; he can restrict him to the limits of quarters or of the post for as long as a week; he cannot however, thus summarily, give him confinement in the guardhouse or fine him any of his pay. As "punishment" it is not very severe. It is a convenient method of meeting with small derelictions. Indeed, it is strongly recommended by higher authority for all "first" offenses and for minor matters, without the necessity of recording against a young soldier the formality of court-martial proceedings. This scheme of company punishment leaves much discretion in the hands of the company commander, and gives him direct control over his men, without the need of calling on higher authority through the medium of formal trial by court-martial. It is not likely to be abused, for a soldier always has the right of appealing beyond his captain to the next higher authority, if he thinks the captain has not made "the punishment fit the crime."

CONVOY. A word which has in the army, as in the navy, a double meaning. Early after its introduction into the English language from the French, a "convoy" became a unit which accompanies and guards something else, and has also been used to describe the unit being guarded. For instance, we find the London Gazette in 1697 saying: "The Chester, with several vessels under her Convoy," and the same paper in 1710 saying: "A great Convoy of Bread came yesterday." In 1763, Colonel Bouquet provided for 90 men to "escort" a "convoy." It may seem silly, but with these two meanings it would be perfectly correct, although of course confusing, to say: "The convoy guarded the convoy." To avoid this confusion we never do use the two similar words in the same sentence. We may say that: "The troops guard the convoy." Or we may say: "A convoy guarded the shipment." Of course, you can call the two of them together a convoy and let it go at that, a thing which is also frequently done.

In the old Indian days, American troops on the western plains frequently guarded long trains of emigrant wagons, or perhaps of wagons bringing food and ammunition from the railway, sometimes fifty or a hundred miles away. This was a particular type of duty, convoy duty, and was given special study. Since the World War, however, this use of the word has dropped into disuse. In France, the front was continuous. No enemies were wandering around behind the lines ready to pop out upon, surprise, and seize a "convoy." (If you know what I mean! That is the trouble with it, it is hard to tell what you do mean.) So the "convoy" of post-war language came to mean only the long line of the trucks or wagons themselves, not the guard. It has been used to designate a group of motor trucks going along together (in the exact sense of the Latin and French origins of the word) and not just ambling separately toward their destination.

Perhaps the land convoy is actually gone forever. Modern bombardment planes are too dangerous for any trucks to travel in close and quick succession down a straight road. Modern "blitzkrieg" penetrations by tanks and armored cars are too dangerous, too. So the convoy is gone, except for peacetime movements changing station.

COOK'S POLICE. This is the same as Kitchen Police, and called throughout the army just plain "K.P."

CORNER POCKET. This is soldier slang for the guard-house, and the guardhouse, as you will see later, is not only the place where the guard stays. It is also the place where the prisoners serving sentence of courts-martial are made to stay. The buck who runs foul of the regulations and is clumped into the guardhouse is just as much out of circulation as the pool ball which goes into the "corner pocket."

CORPORAL. From the French out of the Italian which in earlier romance tongues was *corporal* to designate the head of a small body (*corpus, corporis*, in Latin). See what Digges in 1579 said: "The Corporal is a degree in dignitie above the private souldior." And Barrett in 1598 says: "Of the best approved souldiers to chuse for Caporals." In our army for years, an Infantry corporal commanded a squad of eight men. Now they have changed the Infantry squad to twelve and it is led by a sergeant with a corporal as assistant leader. When a man becomes a corporal, thereafter he does not have to do kitchen police, and many a man thinks as much of that release as he does of the sheen of the new double chevrons on his arm. There is an army rhyme which runs:

> Oh, the General with his epaulets,
>   A leadin' the parade,
> The Colonel and the Adjutant
>   A sportin' of their braid,
> The Major and the Skipper—
>   None of them look so fine
> As a newly minted Corporal
>   Coming down the line.

And, incidentally, there is no such thing as a "corporal's guard," at least in the army. It is used among literary folk as a synonym for a mere handful of men, presumably so small as to be commanded by a corporal. The nearest approach to it is a "relief" of the guard, which is on guard duty, and which is directed on its duties by a corporal. But that is not what is meant by the phrase and we'll have to put it down to pure literature. Soldiers never use it.

CORPS. "Military terminology," says Major Willoughby, "is by no means as clearcut as it should be," and then cites this very word "corps." By the sense and translation of the word out of the Latin, whence we got it, or others got it and handed it over to us, it means just a "body" of troops. Specifically, however, it means a large body of troops of great size—and this in spite of the use of the word "corporal" on the same root, for the leader of the smallest formed body of troops. Corps in all armies means a unit composed of two or more combat divisions. The word was brought into England from France by the Duke of Marlborough. Addison in 1711 in the *Spectator* says: "Our army being divided into two Corps." Shortly before 1700 the French had the word as an "Corps d'Armée" and so it began. And then in spite of this specific meaning, we find frequent use of "corps" in the other, more general sense, to describe the Dental Corps, the Corps of Engineers, the Officers' Reserve Corps.

Of course the term "corps troops" is tied tightly to the combat corps. Those "corps troops" are supply and combat units in addition to those belonging to the divisions which make up the corps, and they stand to the corps the same as "army troops" do to the army. During the Civil War, which seems to have been the first time we really had "corps" long enough to use the term much, we used it that way, briefly, "corps." During the Spanish-American War, however, we imitated the French "Corps d'Armée," officially dubbing our units of comparable character "Army Corps." We did it again in France. But the soldier strives for brevity. And that word "army" was somewhat confusing for there was also a Field Army created, which was called simply "Army." To avoid confusion, therefore, and to attain brevity, the designation in speech practically always, and in written document a good proportion of the time, was shortened to "corps."

CORPS AREA. With the return and the demobilization of the A.E.F. of World War days, all units higher than a combat division went out of existence. The 1920 version of the National Defense Act, however, devised a scheme, based upon the distribution of population in the United States, by which peacetime training could be carried on and wartime forces raised, to recreate a huge battle army again. For this purpose, the old geographical "Departments" of the army were abolished, and their places taken by nine "Corps Areas" in each of which it was planned to have one Regular Army Division, two National Guard Divisions at reduced peace strength, and three skeletonized Organized Reserve Divisions, perpetuating as far as possible the records and traditions of the World War units. Through the economy years, in spite of reductions, the framework was maintained, and much of the army was administered through the headquarters of the various "Corps Areas."

When the European War of 1939 began to get well under way, and America began to rearm and to expand its forces, troops were sent south for extended training. As the training progressed, and Regular Army Divisions were filled to strength, they were organized into corps again, without reference to the particular Corps Areas where they were located or whence they came. There arose much confusion between the Second Corps and the Second Corps Area, separate entities widely distant although bearing similar names. In December of 1940, therefore, the War Department announced that the twenty-year-old Corps Areas should keep their names, and that the field fighting units should be called "Army Corps" instead of just plain "Corps," which had been their popular name in France and still is. In 1942 the various "Corps Areas" were renamed "Service Commands."

COSMOLINE GANG. A derisive term for the Coast Artillery, also called "cosmoline slingers," because their big guns have to be kept constantly coated with a heavy oil or grease called cosmoline as protection against rust or weather. Coast defense guns are so seldom fired and Coast Artillery personnel spend such a large part of their time just greasing and caring for their guns, that they are dubbed the cosmoline gang. The infantryman himself, however, gets acquainted with cosmoline, and very early in his service, when he is just a recruit. The rifle that is issued to him is most likely one that has been in storage for some little time. Before it was put in storage, it was partially dismounted, and all metal parts were smeared with cosmoline. And it is the recruit himself who must learn to take that rifle apart and clean out the cosmoline from barrel and operating metal. He'll get enough of it under his finger nails before he finishes the job, so that he will learn pretty thoroughly just what cosmoline is.

COURTESY. Although the regulations of the army say that all communications between military men will be conducted with courtesy, what is officially called "military courtesy" is not mere social genialty and politeness at all, but rather a rigid set of rules and practices dealing principally with the customs of saluting and rendering of honors. A soldier must salute an officer, the National Colors, the National Anthem, and the bugle calls "To the Colors" and "To the General." The salute must be returned. The junior officer must salute his senior. The salute must be rendered in the prescribed manner. The soldier must say: "Yes, Sir!" and "No, Sir!" and he must stand straight at attention when talking to an officer, although there are many exceptions—as in combat, in offices where contact is continual, and also whenever the officer directs otherwise. He must ask the first sergeant's permission before he speaks to the company commander. Until recently he was required to talk to an officer in the third person: "Sir, Private Jones wishes to ask the Captain for a furlough." Men must jump to attention when an officer enters a room or approaches closely. The sentinel on post must salute by presenting arms, with the rifle. All this is "military courtesy."

COURT-MARTIAL. This is a formal legal court, as its title implies, erected in the martial or military service under full authority of Acts of Congress. It exists so that military men with understanding of military offenses will try offenders themselves, instead of turning them over to civil authorities for trial, as the British used to have to do before the authority for formal punishment by court-martial was conferred on military personnel. The details are all in the "Manual" and it is only necessary here to distinguish between the three kinds of military courts.

A summary court-martial consists of one officer, usually a field officer, who disposes of cases briefly and promptly, although after hearing all witnesses in full.

A special court-martial consists usually of about five officers which acts with more formality, keeps fuller records, and has authority to award greater punishments.

A general court-martial consists usually of about a dozen officers, keeps a full stenographic report of its proceedings, and has authority to award the maximum punishments for specified offenses, except that it cannot try a man for murder or rape in time of peace (that is left to the civil courts) and it cannot cashier an officer without the direct approval of the President.

Courts-martial can try only military personnel, officers, soldiers, army nurses, West Point cadets, and persons serving in the field with the army. Their only authority over civilians is against those caught acting as spies. The oath by which members of the courts-martial are sworn is indicative of their attitude, to "try and determine according to the evidence the matter . . . between the United States of

America and the person to be tried, . . . according to the provisions of rules and articles for the government of the army of the United States, and if any doubt should arise not explained by said articles, then according to [their] conscience, the best of [their] understanding, and the custom of war in like cases." They do this sometimes for long hours and after hours. For years, however, it was forbidden for a court-martial to "sit" after three in the afternoon. The prohibition appearing in the British Articles of War of 1765 was repeated for almost a century and a half in the American Articles. In 1882, General McDowell wrote from San Francisco saying that according to Macaulay the restriction was imposed to insure a sober court and adding: "Officers do not now dine at three o'clock and do not get drunk when they dine, and the restriction has ceased to have any justification." His suggestion was not adopted until 1901.

COWBOYS. The use of the word "cowboys" started among fighting men, drifted into civil life, and has come back home again in a new sense. During the Revolution, it was the title given to a group of Tory guerillas who roamed the region between the lines in Westchester County, and you can find it in many writings of that time and books about that time. Then, about 1877, you find it used to describe the "cowpunchers" of Wyoming whence it spread to general use through the United States in that sense, although later in a slang sense to describe "drugstore" cowboys. Now that the army has become mechanized and it has rough-riding cross country vehicles, the expert who drives that tossing monster of steel is being dubbed the "cowboy" of the group.

CROW TRACKS. A slang name for the chevrons of a noncommissioned officer, which are also called "hooks."

D.D. This is a dishonorable discharge, given a soldier only after conviction by a general court-martial. The holder of such a discharge is prevented from ever again enlisting in the United States Army, Navy, or Marine Corps, and in some states is presumed to have forfeited certain rights of citizenship. It is the custom to sentence to a dishonorable discharge every man who is convicted of an offense carrying with it confinement for more than six months. In cases where the offense is purely military, i.e., disobedience of orders or regulations, as distinguished from civil court offenses like larceny or crimes involving moral turpitude, it is also the custom for the commanding officer to confine the offender but suspend the discharge; and then, by good behavior while confined in the guardhouse or in special disciplinary barracks, the prisoner can earn restoration to duty, return to the army, serve out his enlistment, and eventually secure an honorable discharge.

DAY ROOM. Among officers, and officially, this is known as the "recreation room" of the company, troop, or battery, but among soldiers it is always and everywhere just "the day room." In it are maintained the company pool table, the writing desks, chess and checker sets, magazines, and such books as the company library may contain. Comfortable chairs, nice furniture, attractive lamps, and rugs, render this room more attractive than the bareness of barrack squadrooms where there is little else for a soldier to sit upon than his bunk or his footlocker. Nowadays, the equipment of the "day room" is maintained from what is known as the "company fund," established from commissions on collections made on paydays for little garrison merchandizing activities, and from dividends made on profits by the Post Exchange or "canteen." But it was not ever thus. Back in 1889, a commanding officer wrote to the civilian post trader to say that if he desired to erect a store to serve the needs of the garrison, he must first in connection therewith "provide a room for the use of the enlisted men with pool, billiard, and card tables, and a bowling alley." In the army today the word has been so completely generalized that "day room" is officially used to designate an entire building erected in a training center under the 1940-1941 program.

DEADLINED. When motorized vehicles in this mechanized army are hauled to a garage or repair "park," they are arrayed in rows awaiting their turns for work. They are then said to be "deadlined." The word is spontaneous; its origin is in circumstance rather than in etymology.

DEAD SOLDIER. This is the soldier's term for an empty bottle—a liquor bottle of course. After the World War, it got spread about the country by the American Legion. Although civilians still speak of "killing" a bottle, the "dead soldier" phrase has largely dropped out of civilian circles. The army keeps it.

DECISION. We have in the service a phrase: "seek a decision." It means that a military force tries to engage an enemy so fully and completely that one or the other must lose—completely. Like the "Fifteen Decisive Battles of the World," of course: but the phrase may be applied to other smaller and less important engagements, too.

Yet there is another sense of the word decision, a sense common to the act of every commander who is called upon to decide, not only perhaps if he shall fight, but even how he shall fight. He studies his situation, learns what he can of the enemy, estimates what effect each of several plans of action may have in the face of the enemy's very best possible efforts, balances plan against plan, and then makes a decision. This decision may be to withdraw, to defend, to attack—may even be *how* to withdraw, defend, or attack. This is the officer's greatest responsibility. Once the decision is made, orders are drawn to put it into effect, energy is expended to carry it out, just as when a judge decides a case in a court of law, a whole cohort of bailiffs and marshals and deputies are called in to make the legal decision stick. Robert van Gelder recently in the *New York Times* interviewed a Spanish Civil War officer who illustrated this so well that instead of finding words of my own, I quote:

"In army regulations, the first sentence was the seemingly meaningless one that 'the first duty of the commander is to make decisions.' It seems simple when you read it. You think, 'What is a decision?' Each day I decide what color shoes to wear, what to eat.' But decisions, when the life or death of hundreds of men depend on your decision, that is much else. In Spain I was assigned, as you know, to hold a position. My cowardice told me to draw in my left flank so that if I failed I would be near the French border and the lives of thousands would be saved if we lost. My judgment said perhaps that is right but perhaps it would be better to bend my right flank, though if we lost we would be cut off from safety. That is a decision that hurts all through your body; you cannot sleep, you ache. There is nothing more difficult in life.

"Do you suppose all commanders feel that way? Did Napoleon?

"Napoleon was a victor. When you are a victor, what can hurt you? But when you must fight a long defensive action with no chance of winning, only of holding the enemy off, then with every decision you are in hell. You ache with wanting—but what you want cannot quite be reached. . . .

"Perhaps because the military decision is so difficult to make, that is why when it is made rightly it pays off so well."

DEPLOY. An Old French word *déployer*, with the basic meaning of "unfold," came in Anglo-French, thence into Middle English, and finally became our modern English "display." Over the Channel the old *déployer* developed into the technical military verb *déployer*, meaning to unfold, spread out, or arrange troops in line of battle. Then, later, by actual contact between the two armies, the French *déploy* in the military sense was also taken into the English language, although already in the other form. It remained unchanged as a technical military term. The dictionary quotes Sullivan as saying: "A column is said to 'deploy' when it makes a flank march or unfolds itself, so as to display its front." As long as troops fought shoulder to shoulder in lines of riflemen, the term was specific and proper. Troops went from column into line of battle, from march order to the firing front. Thus in Grant's memoirs you will find him saying: "Carr's division was *deployed* on our right." The French word was adopted unchanged and fixed in our military phraseology. Today it means officially simply to open a column of troops on a broader front.

DETAIL. At the ceremony of guard mount, the sergeant major leaves his post as soon as the men have taken their places and walks along and around the line, counting. Thereafter to the adjutant he says: "Sir, the detail is correct." This use of the word is distinctively military. A soldier is "detailed" on guard, or on fatigue or on kitchen police. The group of men so detailed or assigned, and the list of men on the sheet of paper itself posted on the bulletin board, are both described by the word "detail." Thus one may say: "The detail is posted," or "The detail is up," or "You are on detail," or "Take the detail to the Supply Office." When a company is brought to a stop, one commands: "Company, halt!" If it is a squad, or a section, or a platoon, or a battalion even, you use the proper word there instead of "company." But if it is not a formal unit, however large or small, it is a "detail" and you say: "Detail, halt!" It is often used in jest.

DISCIPLINE. This word has a totally different meaning in military life from that which it long had in civil life.

[70]

To the average citizen it used to mean punishment, jail, standing in the schoolroom corner. In the army it does not mean punishment, the guardhouse, or restriction. Discipline, to the armed services, is closely akin to the word morale, but in a sober sense. Discipline is that state of mind on the part of a unit which enables a commander to get orderly and efficient results. It means making everyone interested in getting the job done in the best way of getting it done. It is the sort of thing you saw in those great temporary associations, like the corn-husking bee, the house-building bee, the trek of emigrant trains across the plains, by which the greatness of America was built on a vast continent that was a wilderness until the American spirit built it in the American way.

When Baron von Steuben framed the first regulations for the American Army and gave the American revolutionists their first thorough training, he openly acknowledged that different methods were needed in the New World. He said that the commanders must have honest interest in their men, and must lead them to soldierly effectiveness by understanding and intelligence. Washington went so far as to issue orders against the use of profanity. Discipline is what makes soldiers, separated from their comrades and their company, still continue their efforts in behalf of the common plan, in spite of the noise and confusion of the battle-field, in spite of danger, difficulty, and death. There is an incident in *The Snare* by Sabatini which will illustrate the point: To Wellington at Lisbon came a dusty, weary, hard-riding horseman with vital messages. "You appear to have ridden hard," the Duke said. "From Alameida in forty-seven hours, my lord, with these from Sir Robert Crauford," was the answer. "That was great horsemanship, Mr. Hamilton." "The urgency was great, my lord!"

It should be added that "discipline" as used in the Constitution of the United States actually means what we now call "training."

DOG ROBBER. A soldier who does extra-duty work for extra pay for officers. This at least is what the soldiers call him. The officers call him a "striker" and the words are both American inventions. The Englishman calls him a "batman," but the A.E.F. refused to adopt that bit of British speech. Of the captain's "dog robber," Westbrook Pegler has said that he "gets the leavings from the table and apparel of the head man and a chance to slip in an occasional needle load of company gossip and advice."

DOG TAG. The identification tag worn by soldiers in the American Army, on a string around the neck, like a license tag on a dog collar. It is an official adoption of a custom instituted by citizen soldiers of the American Civil War, who before a big battle used to pin on their uniforms pieces of paper on which they had written their names. If a man thus labels himself, it is all right with him. But the moment the government does it, and gives him a "number" besides, the device becomes a "dog tag."

DOUGHBOY. An infantryman, not—as it is frequently misused to signify—any soldier. The less you know about the origin of this term, the easier it is to announce it. In fact there are many explanations, and all plausible. In the British Navy for many years it was used to describe a sort of dumpling. But the question is how this got transferred ashore as a label for foot soldiers. It is said that the buttons on their uniforms looked like these "doughnuts" or "dumplings" and so were called doughboys, and the wearers also doughboys after them. Custer in 1867, a cavalryman, declared himself "not a doughboy." It is also said that Crauford's Light Division in the Peninsular War, famous for its marching qualities, once ground its own wheat into flour so much that they christened the place "Doughboy Hill" and themselves "Doughboys" and the title stuck. It is also said to come from the word "dobe" to indicate a certain kind of mud, used to build huts in the West and also difficult to march in, and to have grown up as a change from an Infantry nickname of longer tradition "mud crushers" in places where the mud they crushed when they marched was "dobe" mud. It is further said that infantrymen who used to clean their white trimmings with a white pipe clay found that the rain made the pipe clay run and look like dough, and hence the epithet "doughboys." An old officer whose service began in the 1870's says that although he often heard the term "mud crushers" used in those days, he never heard "doughboy" until very recent times. Among these stories of the origin of the word you may choose for yourself. Certain it is that the term was widely popularized early in the twentieth century. It is an epithet accepted without derision, even proudly, and is in general use among all infantrymen.

DRAGOON. Mounted soldiers trained to fight as infantry, as for example two of the present Cavalry regiments in the American Army which were previously called the "Mounted Riflemen" and the "Dragoons." Originally there was a close distinction between the cavalryman and the dragoon. Macaulay say: "The dragoon has since become a mere horse soldier. But in the seventeenth century he was a foot soldier, who used a horse only in order to arrive with more speed at the place where military service was to be performed." For example, Defoe in the early eighteenth century thus described a cavalry unit in battle: "They lost most of their horses and, turning dragoons, they lined the hedges."

These early dragoons, then, who were carefully distinguished from the "Horse," were essentially trained to fight on foot. Lacey in 1672 said "Dragoons are but foot on horseback." Then, during the next hundred years or so, with the "shock action" of Marlborough's, Frederick's, and Napoleon's cavalry, there was an increase in mounted action. It became later the practice to train all cavalry to fight both on foot and mounted. This has been the American practice, illustrated best perhaps in Indian campaigns on the western plains, and in Wilson's cavalry campaign in Alabama, and in Forrest's roving fights. All horse cavalry today, then, are really "dragoons," although not so called.

The origin of the name "dragoon" is somewhat obscure. Sometimes it is said that this species of force took its name from a short carbine, sixteen inches long with a large bore (called a "dragoon" and carried by the original dragoons raised by Marshal Brissac in 1554), on the muzzle of which was a small sculptured dragon. Others say the name originated in a standard or ensign, so called because it was decorated with a dragon painted or embroidered upon it, or because it consisted (as in the Bayeux Tapestry) of a figure of a dragon upon a staff. The standard was a "dragon" just as those of Rome and of Napoleon were "eagles." In any case, the word, which appears in most continental languages, Dutch, Danish, German, Swedish, French, Italian, and Spanish, is very old in English. It appears in poetry and ballads, as "draggoons, and horse, and foot" and as "dragoons and foot."

DRESS. Not satisfied with the several meanings of this word in ordinary civil life, the army has taken it over and given it two special meanings of its own. One of these is connected with clothes—"dress uniforms" which are part of the "blues" of the army. The other meaning has to do with drill. Men of a unit are supposed to be on line, but if the line is too crooked, the commander says: "Right dress!" and everyone by raising arms or elbow (whether they are "extended" or "closed"—see the drill books!) corrects the interval between himself and his neighbor in ranks. He also turns his head to the right and corrects the "dress" of the line, which is to say simply that he corrects his own position on the line. The word appears as early as 1746, and is found in America in Duane's drill regulations of 1813.

DRILL AND BE DAMNED. A slang phrase for that portion of Infantry training designated at service schools and in manuals as "Drill and Command"—usually employing the "cadence" system, in which all men in ranks at a given signal shout aloud in cadence the proper words of command and then execute the movement themselves. This "cadence" system was promoted during the World War by Major Bernard Lenz and proved to be very successful for the rapid instruction of many men. Lenz's system has recently been revised to fit the new drill regulations, and has been frequently used in the training of our new recruits.

DRIVE IT INTO THE HANGAR. This is the way fellow aviators tell one who is "bunk flying" to change the subject.

DRY RUN. This term originated on the target range. At rapid fire it was customary to have the targets exposed the regulation time and to have the shooters go through the motions of shooting their ten shots, without ammunition, and for practice only. Since the World War this has universally been called a dry run. It is used in machine gun practice, too, but it has no relation to the fact that since nothing is going to be shot out of the barrel, it is not necessary to have cooling water in the water jacket. It was used by riflemen before we had any water-cooled machine guns in our army. From this target range usage, the word has spread to be employed to describe any "dress rehearsal" as for a parade, a demonstration, or presentation of any sort. "This is just a dry run," means that this is just practice and does not count.

DUCKS. This is straight soldier slang, whichever meaning you give to the word. In the 143rd Infantry in the last war, I am informed, they called the new men shipped to them as replacements by the name of "ducks" for it seems they were told not to call them "conscripts" and the words "draftee" and "selectee" were too formal for soldier speech. They called them "ducks" throughout that regiment, because these men had been inDUCted into the service. The word, however, was not widely used. Indeed, it is not definitely known to have been employed outside of that single regiment, and would not have been included here were it not so apt an illustration of the makings of an oral language on the lips of military men, and so pat an evasion of exact and legal phraseology. The word "ducks" had, nevertheless, in many circles another use and meaning in the service and still has. The soldier calls "ducks" those small sums of money which are deDUCted from his monthly pay for minor charges like barber fees, canteen checks, tailor payments, and so on and on until often there is little left. These "ducks"— from the soldier's habit of living so often "jawbone" or on credit—sometimes eat up a very substantial part of his pay, unless a wise company commander puts reasonable limits to their variety and amounts.

DUCKS. Air Corps men have another meaning for "ducks." They give the name to amphibian airplanes, capable of landing on water or ground. This true metaphor was widely used by army aviators in the 1920's, almost died out for a decade, and has returned to general use with the procurement of many planes of this type.

DUD. A rare bit of American and English slang, which, applied to a person, means a useless chap or a scarecrow. In 1840, Carlyle dubbed a caller "a wretched dud." In the army this old slang word has been taken over to describe artillery shell or an aerial bomb which does not explode—a shell that is useless or worthless for the purpose for which it was hurled—a shell perhaps that is without any more real value than a scarecrow, if you scare easily. It may give you a good fright, make you duck your head, or leap for a hollow spot, but it does you no real harm. In France, *The Stars and Stripes*, in its issue of April 12, 1918, said that Americans did not take over many English expressions and declared: "The only really permanent acquisition thus far from the British is the expressive 'dud.' "

DUMP. This, in the army, is not the place where you get rid of things you do not want, but the place where you get the things you do want. When a force is campaigning against the enemy and supplies are brought forward for distribution successively to lower units, they are temporarily held and sorted out (or "issued") at an intermediate depot. Because storage under cover was often not available in France, the American Army began talking about the "Division Dump" and the "Corps Dump"—for the ammunition and the boxed rations and other things were often just dumped on the open ground. You can talk about "ammunition distributing points" and "supply points" all you please. But, so long as much of this material is likely to be put on the ground, the soldier is going to call it a "dump" no matter what the books and the supply officers call it.

DUST INSPECTOR. When a soldier turns out for inspection, he has his hair cut, his buttons polished, his face shaved, his clothes pressed, his shoes shined, and his rifle or other weapon cleaned to the last word, until he and his clothing and accouterments simply gleam. But there is dust in all atmosphere, and some of it will inevitably settle on one or more of the highly glistening military surfaces. In such circumstances, the soldier gives the sneering epithet "dust inspector" to any officer who is exceptionally fault finding and will—for lack of any real defect—note and record the light recent dust which may have collected on boots or rifle stock or barrel after these have been carefully cleaned just before inspection. Going through barracks, inspectors of this sort have even been known to take framed pictures down off the walls of the "day room" and point with serious mien to flecks of dust on the cord or nail from which they hung.

EAGLES. Upon each of his shoulder straps, to designate his rank in the army, a colonel wears a silver eagle. Getting promoted is thus said to be "getting your eagles." The soldier is likely to refer to them as "buzzards" and even to the colonel as "the old buzzard," but they are really eagles. A change in uniform effected June 11, 1832, established them as the insignia for the colonel in the United States Army. They are typically American, other forces not using them. A "full" colonel is often called an "eagle colonel."

ENFIELD. When America entered the World War, the standard army rifle was the excellent Springfield. This requires time and many special tools to make properly. The

Richard Hurn

Engineers

army expanded rapidly from a mere hundred thousand to four million, and the arsenals could not produce Springfields fast enough. This situation was foreseen. So they took the design of a British Enfield rifle, which had been under manufacture in large quantities, made relatively slight changes in it, bored it to take the standard American bullet and issued it to many divisions of the World War Army. Officially it is the "U. S. Magazine Rifle, Calibre .30, Model 1917" but it is always familiarly called the American Enfield or just the Enfield. The soldiers did not like it. Every time they got a chance, they picked up a Springfield dropped by some other soldier killed or wounded and threw away their Enfields. Although during the last twenty years many of these Enfield's have been sold by the government, there are still many in storage, awaiting the day when a rifle good for short range shooting only will be needed. Some of them were issued to the "State" Guards of the various States in 1941. Some were turned over to the British Home Guards, who mistakenly call them "Springfield's."

ENGINEERS. This word, which now has many uses, was taken over, only slightly changed, from the French. At first all engineers were military engineers, like Vauban, who built canals and roads for France, as well as splendid ramparts and forts. Then came those who built lighthouses, roads, and bridges for commerce, and in 1793 were termed "civil" engineers for their works were civil rather than military. Nowadays, the military engineer tends to water supply plants for the troops, construction of big cantonments in the theater of operations, and accurate mapping and map reproduction, as well as repair of bridges and roads, construction of permanent fortifications, and the throwing of temporary pontoon bridges across a stream in the face of an enemy. They may also be combat troops with some of the most difficult and dangerous tasks of modern war. Although the British call them "sappers" from their early use in digging "saps" toward the walls of a besieged fortress, the American Army does not use the term.

ESCORT WAGON. The combat wagon was that in which the mules hauled ammunition and other items designed for the combat use of an organization. It was strong and heavy and certainly took four husky Missouri mules to haul it. Its load was heavy. Then if we go back to Indian days we shall find that there was devised a lighter wagon which was employed for carrying the armed escort of supply columns of the wagon train. The escort of soldiers, armed against attack by the red men, rode in this wagon; so it was called the "escort wagon." In time as the need for the extraordinary weight and strength of the combat wagon gradually diminished with the betterment of roads and routes, the "escort wagon" began to be used as a "combat wagon" and in recent years the escort wagon has been practically the only one ever used. So now the combat wagon is an escort wagon, or the escort wagon is a combat wagon—whichever way you want to say it—at least in the few organizations and posts that still have animal drawn transportation.

EXPERT. The Continental Congress in a resolution of June 14, 1775, called for the formation of companies of "Expert Riflemen." This was a general, descriptive use of the word. The army now has a more special and official use. Today an expert rifleman is an individual who has achieved on the target range a certain score in record firing, e.g., 306 out of a possible 350. The rating he gets and the badge he wears mean that he is an accurate shot of the highest class, not that he is an expert technician or mechanic on the weapon. There are also Expert Machine Gunners,

Expert Gunners with tank and antitank guns, mortars, and other weapons, and also Pistol Experts, all determined by firing high record scores in annual target practice.

EYEWASH. This is a common army term for the result of prettying up a post, to appeal to the eye of an inspecting officer. A piece of ground may be clean; but if you run a rake over it and rough the surface, it looks as if some work had been done on it. It is eyewash. The British *Notes and Queries*, in January 1919, said it was also a British Army term, and indicated that it implied some deceit, meant to cover actual faults with surface prettiness. That is not a part of its meaning in the American Army. If a regiment goes out on field maneuvers, and instead of studying tactics, sees that its tent pins are whitewashed, and all lined up as if with a transit, it is "eyewash." The worst of it is that too many generals have been more interested in "eyewash" than in anything else. The army even has the tale of an automobile which won a transportation competition, because it excelled all others in the shine of its hub caps, the gleam of its varnished sides, the contrasts of its flecks of red paint on protruding bolts—and its engine was such a wreck that it could not run. That is one of the fatal results of "eyewash." And there is a story of last year that when a shipment of automobiles arrived at one place carefully painted with a dull finish so they would not shine in the rays of a tropical sun, a general directed that they be waxed and rubbed to a high polish.

FALL IN. This and its, later (oh, much later!), complement "Fall out" are actual words of command. They mean respectively to get into line and to get out of line. But "fall out" is also used as short for "fall in outside"—when the top kick orders the company out of barracks for a formation.

[83]

FATIGUE. Labor, as distinguished from military drill. It may be heaving packing cases around for the quartermaster; it may be mowing the grass on the parade ground; it may be whitewashing fences, posts, or garbage cans. It is not popular with troops, who would rather drill or have strictly military training. Like the boy who said: "I jin'd the army 'cause I wants to dreel. An' ain't dreeled yet!"

This is "Fatigue" call:

FATIGUES. One-piece unionalls, or separate trousers and loose coats, usually of tan or blue denim goods, which make up the working costume for men engaged in "fatigue."

FIELD. Again, as always, the army abbreviates and says "The Field" when referring to the Field Artillery, which the gunners call "artirry." The Field Artillery operates the "75," the "105," the "155," and 8-inch guns; and all the heavier guns, railroad, tractor-drawn, and those in fixed fortifications, are operated by the Coast Artillery. (See also "In the Field.")

FIELD MUSIC. This is not just music played in a field, rather than in a band stand. The full complement of a regiment's musicians are a band. Drums and bugles alone are called the "field music"; indeed, bugles alone fulfill all requirements of "field music." The big band is left behind in barracks. The massed bugles play the important calls together, and this is "field music"—the group of buglers themselves, not the music they play.

FIELD OFFICER. There are three main classifications of officers in the army according to their grades. (1) General officers are those of the grade of brigadier general, major general, lieutenant general, and general. (2) Field officers are the colonels, lieutenant colonels, and majors. (3) Company officers are the captains, first lieutenants, and second lieutenants.

FILE. In drill ground terminology a "file" consists of men one behind the other, "in single file" as the saying goes, or "in Indian file." Thus we have the phrase, "cover in file," which means that each man in rear should adjust his position so that he stands directly behind the man in front of him. "File" has, however, another peculiar meaning when used in relation to commissioned officers. All officers are arranged in order of seniority, and there are said to be so many "files" between the relative rank of one and of another, and an officer reduced in rank is said to have lost such-and-such a number of files. It is almost as if all officers of the army were marching toward the top in single file. But the custom has arisen of speaking of an individual officer as a "file," and it is common to hear an elderly one spoken of as "a good old file." But a man called a "bad" file, is not bad. A "bad" file is a young man who has higher rank on the promotion list than older officers who will reach the age of retirement before he does. His relative youth, and the fact that in the years to come he will not retire for age before they do and so will not thus create a vacancy for them to step into, cause him to be spoken of as a "bad" file. This matter of files was, until 1940, the entire basis of promotion in the grades from second lieutenant to include colonel. Now permanent promotion in the commissioned grades is based upon years of service.

FIRST CALL. This is a piece of bugle music sounded as a warning, usually fifteen minutes before the time for troops to form for any duty. Thus, ahead of "Drill Call," and ahead of "Fatigue Call," and ahead of "Assembly" for parades, they blow "First Call." It is worth noticing that this, not "Reveille," is the earliest call sounded in the morning and the one which actually wakes the soldiers up.

FLAGPOLE TRANSPORTATION. When a soldier's enlistment expires and he is discharged, he is given all pay due him, and in addition a sum of money calculated as sufficient to pay for his transportation to his home. But soldiers often re-enlist at the same army post where they are discharged. Consequently, at the end of this second enlistment, which commenced at the army post, the government is obliged to pay a soldier only what is necessary to carry him to the place where he enlisted the second time, that is, to the very post where he is serving, and consequently owes him and pays him no travel pay at all. This is what soldiers call "flagpole transportation." When they say they get it, they mean they get none.

FLIGHT. Administratively, military airplanes are grouped into squadrons. There was a time when, if they were not flying alone, they usually flew in threes, although sometimes in larger numbers. A set of nine or fewer planes is officially in the American Army a "flight."

FOOT DIVISIONS. Elsewhere a distinction is made, on the basis of established army policy, between "mechanized" troops and "motorized" troops. The first are those which fight on their vehicles, like tanks, and armored cars, and reconnaissance vehicles. The old distinction, however, between "motorized" troops and those whose vehicles are animal-drawn, has begun to break down. The tendency today is to speak of "foot troops" to designate those who march, however their ammunition, baggage, and rations may be hauled; and to speak of "motorized troops" to designate those who do not march but are hauled in motor busses or trucks, even though the men actually get out of the vehicles later and fight on foot. These "motorized divisions" are used for swift moves to reinforce threatened portions of the line, or to follow up and hold ground which tanks and armored forces ("mechanized" troops) have seized. They are essentially mobile forces. The "foot divisions" on the other hand comprise the general fighting force to push home in detail the further conquests of all arms. The footslogger has not been brushed from the battlefield by the gasoline-powered motor car. In the 1940 fighting in Belgium and France, the largest portion of the victorious German Army was composed of infantrymen who marched on foot, were hardened to march under full loads as much as thirty miles a day for several days, and did march at that rate, to consolidate the conquests. In 1941 in Greece, when the German "panzer" divisions were stopped, the mountain troops who drove in and around the British were foot troops.

FOOTLOCKER. A locker in a gymnasium, country club, or machine shop, is a tall steel closet which stands against a wall. In the army, this type is called a "wall locker," and in permanent barracks each soldier has one. The army "footlocker" is not a locker at all, in the ordinary sense, but rather is simply a small trunk, about the size of a steamer trunk, usually painted olive drab and decorated with the company and regimental letter and number and with the divisional insignia. It is called a "footlocker" because it habitually stands in barracks at the foot of the soldier's bed. Civilians call it an "army trunk" because it is of distinctive design and size. Soldiers call it not only a "footlocker" but also a "trunk locker." Why it is not possible to call it just a "trunk" no one seems to know.

FOOTSLOGGER. An infantryman, who moves on foot by marching. Kipling's famous "Boots, Boots, Boots" rhyme of the South Africa War has the phrase: "We go foot-slog-slog-slog-sloggin' over Africa!"

FORMATION. The assembling of units in ranks, and thus by analogy any specified duty at which one must be present. In the British Army they have a "parade" for this or that, while we have a "formation."

FORT. By definition and origin in the French word meaning strong, a fort is a defended place. It should be provided with permanent entrenchment, embankment, palisade, or other physical means of protection against enemy fire. But in the United States the word is not nearly so definite. It is used, of course, concerning coast defense installations, like Fort Hamilton, and Fort Hancock. Also concerning Fort Monroe, which is familiarly called Fortress Monroe by civilians. And also concerning old-time masonry forts like Fort Jay and Fort Wadsworth. But it is also used to describe places like Fort Ethan Allen and Fort Monmouth, where no "fort" in the true sense of the word exists or has ever existed, or where no one ever intends to put one. A "fort" to the army, then, is an official part of the name of any military station where combat troops are stationed, provided it is a permanent post and not merely a temporary one.

There is some chance that the practice of naming all these places forts is based on the constitutional provision that the federal government could assume exclusive jurisdiction over ground within the limits of the several states that was used as a "fort" or "arsenal"—but this theory breaks down when we see Jefferson Barracks and Plattsburg Barracks so named without any word "fort" attached to them at all. During the World War, the various cantonments erected for emergency training were called Camp Bragg, Camp Devens, Camp Meade, etc. After the War such of them as were retained and transformed into permanent stations for regular troops were renamed Fort Bragg, Fort Devens, Fort Meade, etc. A few others, still held, but not permanently garrisoned, retain their World War designations as "camps," for they are used only for summer training seasons. This business of calling almost every post a "fort" has its misunderstandings. Fort Sheridan has no ramparts or embankments. A visiting lady asked an officer: "Where is the fort?" He replied: "I am sorry, madam, but that is a military secret."

FOURRAGÈRE. Straight from the French during the World War. A braided cord with a metal point worn over the left shoulder. An honorable decoration granted by the French government to organizations which have distinguished themselves in battle, to be worn by all members of those organizations. It comes in the colors of the Croix de Guerre, the Medaille Militaire, and the Legion of Honor, and of the Tri-Color of France. "Fourragère" is simply a general organization decoration. In form it is an aiguilette, for which we have already given a long explanation above. When it is awarded to a unit, all members put it on. New recruits are privileged to wear it while they are in that unit, even though they were not in the unit when it was "decorated." Men who were in the unit when it was decorated may continue to wear it as an individual decoration, even after they have been transferred elsewhere.

FOXHOLE. Officially this was long called a "skirmisher's trench." But beginning on campaign in France and continuing to this day, the soldier speaks of it as a "foxhole"— which covers all varieties of hasty field entrenchment whether for a man kneeling, sitting, standing, or prone. This very unofficial character of the name simplifies language, making one term easily cover all forms without being too particular about which form. Now, for protection against aerial bombs he digs narrower and deeper and calls it a "slit trench."

FRONT AND CENTER. At the ceremony of Guard Mount, the adjutant at one point commands: "Officers and noncommissioned officers, front and center, *March!*" They move out toward him, halt, and await instructions. From a part of a precise ceremony, the phrase "Front and Center" has slipped into familiar speech. The first sergeant may stick his head out of the orderly room door and shout: "Jones, front and center!" and Jones comes. Or a group of soldiers standing talking may hail a passing friend to join them by calling: "Front and Center!"

FULL FIELD. The full equipment of a soldier ready for field service, including pack, ammunition, rations, weapons, blankets—even down to the toothbrush. Orders for important inspections usually say: "Full field equipment, less ammunition." It is bad enough to stand at attention wearing all this weight, waiting for an inspector to come around. It is worse carrying it for fifteen miles, especially the fourteenth and fifteenth miles. Every trained soldier can do it, but nowadays the bulk of the pack is carried in a unit truck.

FURLOUGH. An authorized absence from duty granted to a soldier by his commanding officer, for a longer period and somewhat more formally than a "pass." For officers, the term is not used; instead their privilege of this sort is called a "leave of absence" or a "leave." A man is said to ask for a furlough, to be granted a furlough, to go on furlough, to be on furlough, and the very sheet of paper by which written permission is granted him is itself called a furlough. The word came into English from the Dutch, *verlof*, during the many wars waged in Flanders by British troops. Ben Jonson in 1625, a veteran of those conflicts, spoke of the preciousness of "a Low-Countrey vorloffe." In 1707 Farquhar has a grenadier "absent on furlow" and in 1804 Wellington uses the form "furlough."

G. I. "Galvanized iron." The big ash cans of the army are always spoken of as " G. I. cans" and the water buckets or pails as "G. I. buckets." Never just "cans" and "buckets." Because of their size, heavy artillery shells during the World War were also dubbed both "ash cans" and "G. I. cans," just as the French soldiers called those same shells in their own language *marmites* (i.e., large containers used for carrying food) for the same reason. We have recently heard, at the Tennessee maneuvers, the phrase "galloping G. I. can" used for tanks, but its permanence is questionable.

G. I. Originally "general issue," as when used in official lists of supplies, as in designating the "G. I. soap," which is the standard yellow soap used by the army for washing dishes and clothes. It has been, in popular soldier speech, extended in its use when referring to articles to which it is not officially applied. For instance, one might speak of food in the mess hall as "G. I. Food." Its most interesting extension is its use to describe the girls whom kindly chaperones bring to an army dance from neighboring towns to help entertain the soldiers; they call them "G. I. girls," and the dance a "G. I. Hop." The soldier calls peace-time maneuvers a "G. I. War" and a close military haircut a "G. I. haircut." This use of these initials as an abbreviation of "government issue" is fairly recent. The Old Army—before the present war—never spoke of a "G. I. soldier."

G-2. A military staff is divided into administrative, information, operations, and supply sections, numbered respectively G-1, G-2, G-3, and G-4. Others are more likely to be called by their names. But the official name for G-2 people is "intelligence" and the soldier does not like to compliment anyone. So he clings to the shorter and less complimentary form: "G-2." G-2 duties are to get information of the enemy, or have others get it, to get good maps of the area of operations, and to try to figure out what the enemy is going to do, or can do next. A classic example of good G-2 work was that performed by Americans in France, when they predicted almost to the mile of front and to the day, the place and time of the great German offensive of March 1918. In addition, we should add that "G-2" is by analogy popularly applied to any sort of sneaking and snooping around into other people's business.

GARRISON PRISONER. When a soldier has been found guilty by court-martial of what is after all a purely military offense, he serves out his sentence at the same post where his regiment is stationed. He is thus called a "garrison prisoner" to distinguish him from a "general prisoner," who has been sentenced to a long term and a dishonorable discharge, and is usually soon transferred elsewhere. The sentence of a garrison prisoner can be commuted by the post or garrison commander; that of a general prisoner can be altered only by that higher authority which appointed the general court-martial which tried him. The distinction is therefore technical and legal, as well as descriptive and geographical.

GENERAL. Originally a march piece played by field music, i.e., by bugles and drums, as warning of a general assembly. Wellington in 1803 said: "The generale was beat at half-past four; the assembly at half-past five, and we marched immediately after." Nowadays its warning functions have been taken over by "First Call" and "The General" has another specific use, dating according to veterans in our service as far back as the Civil War. It is commonly played when troops are in the field in tents or shelter tents and are due to move out. Equipment and supplies are packed, guy ropes are loosened, and the procedure is for all to stand by and to have all tents fall to the ground simultaneously on the last note of the music of "The General."

GENERAL OFFICER. A general officer is one of a group who bear as parts of their titles the word "general," that is: brigadier general, major general, lieutenant general, and general. Sometimes one of these is spoken of just as a "general" and sometimes as a "general officer." The latter phrase dates from as early as 1681; and became common in the British Army. It was used in 1759 by Sir Jeffrey Amherst over here in the American colonies; and later by our own forces during the Revolution. We got it from the British, of course, and although we do not use it very much, we do use it, generally when the main emphasis is being placed upon the perquisites of rank rather than upon the personality of the individual general himself. For instance, the sentinel shouts: "Turn out the Guard, general officer!"

GENERAL ORDER. An official order from a military headquarters of interest to an entire command—promulgating, that is, general principles or instructions. They are distinguished from special orders, in that the latter pertain usually to assignments and movements of single individuals. They are usually numbered serially at each headquarters each calendar year. This type of general order is officially and almost universally abbreviated in writing as "G.O."—although only rarely in speech—and "G.O." is not used for the "general orders" of a sentinel, or for "general officer" which the initials would fit just as well.

GENERAL ORDERS OF A SENTINEL. A set of all-inclusive instructions to sentinels on posts of the guard which are uniform throughout the army and are learned and followed by all members of military guards. Their general character can be exemplified by quoting the phraseology of three of them:

"To take charge of this post and all government property in view."
"To walk my post in a military manner, keeping always on the alert and observing everything that takes place within sight or hearing."
"To quit my post only when properly relieved."

GIG. From West Point comes this word common in cadet slang, meaning an official report of a minor delinquency; anything from lateness at a formation to a fleck of dust on a button or a grain of sand on a shoe. Graduates brought it into the army, but it did not last long. It was soon almost completely replaced by the more vivid and figurative "skin." At the reception and training centers of the mobilized army of 1940, it was suddenly resurrected. Currently it now has a more specific meaning than the older phrase. A "gig" is now confinement to the post for a week-end as punishment, although sometimes for a longer period.

GOLD BRICK. There is no implication of dishonest dealing in the army's employment of this phrase. It is used, and not too invidiously—in fact used so frequently as almost to be generally acceptable—to refer to a soldier on special duty away from troop drills, as for example, a man in the military police detachment, in the Post Exchange, or officially absent from "straight duty" with any line organization. All headquarters clerks are likely to hear themselves called "gold bricks." Also the word is used to apply to being lazy on even a legitimate task, as when a man is told to "quit goldbricking on that job."

GOLDFISH. Although it is said that in some places, the title of "goldfish" is applied to fried carrots on the mess table, this use is restricted and not very venerable. It has stood for very many soldier-generations of usage as "goldfish" to refer to canned salmon, a very frequent way of giving soldiers a balanced diet with certain necessary vitamins which seems to have been common to all modern armies since canned foods were invented. It is one of the few phrases of descriptive slang which was equally well entrenched in the British and the American Armies when last they fought in France.

GRAVEL CRUSHERS. Long used in both the British and American Armies, particularly by mounted troops, to describe infantrymen, from their constant need of marching on gravel roads. Sometimes it is "gravel agitators."

GRENADIER. Originally a soldier who threw grenades, often in addition to carrying a rifle or musket. Sir John Evelyn in 1678 said: "Now were brought into service a new sort of soldiers call'd Granadiers, who were dextrous in flinging hand granados." The use of grenades by soldiers was continued only a short time, but in that time they acquired a distinctive name as well as distinctive dress, as evidenced by the old song, "The British Grenadiers":

> Then let us crown a bumper
> And drink a health to those
> Who carry cups and pouches,
> And wear the loupéd clothes.

Increased effective range of rifles put the grenade out of fashion, although the name "grenadiers" remained, being applied to the tallest men in each battalion who were grouped into a separate company and entitled "Grenadier Company," which was only a title. In America, General Charles Lee in 1776 said: "I have formed two companies of grenadiers in each regiment." Smollett's *Roderick Random* in 1748 spoke of belonging to a grenadier company at Dettingen. Napoleon used the designation for special troops in whom there was great confidence. They marched at the head of a column, were used on important posts guarding vital bridges, or employed to sustain threatened pickets. It got to be merely a title of honor.

During the World War, when hand grenades were revived again for trench warfare and its close combat work, after the example of Port Arthur, the name grenadiers was also revived for special groups of bombers. But in the British Army its use was checked by official order in deference to the "Grenadier Guards," and in the American Army the doctrine of all-around training and of open warfare methods prevented such specialization as would warrant its full adoption here, although Americans from 1917 to date have stuck to the word "grenade" and do not use "bomb" at all except in connection with airplanes.

GROUP. During the World War, when artillery batteries were temporarily grouped together, the French called them a "group" rather than the formal "battalion." We used the word occasionally there with the same meaning. Since then it has become an official Field Artillery term. It has also been adopted in our service by the aviators, to whom a "group" is now the official term for a group of air squadrons, and corresponds in general to the ground soldiers' "regiment." Special temporarily formed units are also given the term, as "reconnaissance groups," formed on Carolina maneuvers by assembling motors from various companies.

GUARD THE FLAGPOLE. For minor offenses, a soldier is often given the light punishment of being restricted to the post or camp. So, such a soldier is in garrison slang said to be "guarding the flag pole." The phrase originated in large army reservations and also in temporary camps without official reservation limits, where the limits for restricted persons were likely to be set at a mile from the flagpole.

GUARDHOUSE. This is, briefly, the house where the guard stays. It also almost always happens to be the place where "garrison prisoners" are confined when they serve short sentences. So, although the term is frequently used to refer to the building in its first sense as the post of the guard, it is also used as a synonym for punishment: "He was put in the guardhouse." In its strict, original meaning as the house where the guard stays, the recent *Dictionary of American English* finds it used in Georgia in 1735 and in New Hampshire in 1775, in the latter as a house "for the accommodation of the Soldiers" detailed to guard the coasts. Such a place was in the seventeenth century called a "court of guard."

GUARDHOUSE LAWYER. A soldier who pretends to a knowledge of regulations to such an extent as to stand as a prime example of the adage: "A little knowledge is a dangerous thing." He is constantly citing regulations or the court-martial manual to show why this or that can or cannot be done, or why the officers are not right in treating him as they do. He is free with his advice. He is an amateur lawyer, and as such a jargonist. He is called a "guardhouse lawyer" because usually the course of action he recommends will lead him or those to whom he gives advice, straight to the guardhouse. In the army, reasons why a thing cannot be done have a habit of vanishing into thin air.

GUMSHOES. Members of the military police. The nickname has been known in civil life for many decades to refer to detectives who are supposed to go sneaking around on rubber soles. Military police, patrolling the cafes and watching behavior on the streets, do not do that; but the soldier in his usually uncomplimentary fashion has forthwith named them "The Gumshoes." The title may not fit, but it sticks. It is also used to designate those who do G-2 work.

GUN. In the army this word is used only to refer to artillery pieces, and not even to all of them. The shoulder weapon is never a gun but a "rifle" or a "piece." The origin of the name is obscure, but one of the many explanations is that "Gunnhildr" was the pet feminine name given ballistae and early cannon in the first days of gunpowder in the West, just as the musket was later called "Brown Bess." But the word "gun" is English for many centuries, going far back to a Middle English "gunne." In the artillery they have "guns" and "howitzers," and "mortars," the first, flat-trajectory weapons of high velocity, the others, those which shoot projectiles high into the air to hit the target with a downward rush. The mortar or the howitzer was often installed in coast defense forts, because it was considered more effective to have shot fall downward onto the deck of a ship instead of hitting its side. They are used also in modern field warfare, because "howitzers" can be well hidden behind steep slopes and shoot high over the crests without their flashes betraying their positions to the enemy.

GUNNER. The soldier who aims the gun in the artillery. The others at the piece are officially called "cannoneers." But this distinction is modern and official. Barrett in 1598 said: "With euery peece of Ordnance there ought to go two or three Gunners." It is customary in popular speech to refer to all artillerymen as "gunners" and to their unit as "The Gunners." This label is also common in England.

H.E. The high explosive shell is thus distinguished in the Artillery from gas shells and shrapnel.

H-hour. The fixed hour for an attack. An attack under the conditions under which the A.E.F. fought in France was carefully prepared well in advance. Its exact date and hour were not announced long beforehand, because they wanted to keep them secret. And so, in all the orders, they were spoken of as H-hour and D-day. When all preparations were made, word would come down by officer courier confidentially that H-hour was to be 5:00 a.m. and D-day September 26, for example. The British spoke of it as "zero hour" from which successive stages of the advance, and successive timings of the creeping artillery barrage were calculated. But the Americans stuck to H-hour. They still use it.

HANDSHAKER. From the act of shaking hands as a symbol of friendship came this word to describe the man who shakes hands to become a friend, and thus also who makes friendly efforts to ingratiate himself. In the American Army, where it is an old word, it is used only to describe a person who tries to curry favor with a military superior, and the millions of the A.E.F. came home in 1919 to fix it permanently in general speech in the United States. Handshaking is of course common in civil life; but the formal army greeting is the salute, so the "handshaker" title is in military circles *per se* invidious.

HASH-MARK. The straight stripes on the soldier's left sleeve which indicate length of service. Originally these hashmarks were called "chevrons" and were instituted in the American Army in the following manner. An order issued at West Point, June 17, 1782, said: "The commandant thinks proper to direct that each noncommissioned officer and private who has served four years in the continental regiment, shall be entitled to wear one stripe of white tape, on the left sleeve of his regimental coat, which shall extend from seam to seam." Shortly afterwards, an order issued at Newburgh, August 7, 1782, reduced the number of years from four to three. As time went on, many changes were made in the uniform of American armies; portions were discarded or altered; the service chevron (which is now not in the form of a "chevron" at all) is the only item of the uniform today which had its origin in the days of the Revolution. When you see a soldier with a long row of them from elbow to cuff, you stand in the presence of an "old soldier" indeed—one who has eaten army hash for many years.

HEAD BUCKET. As you will see later, soldiers of 1917-1918 called the steel helmets "tin hats" and the name continued through the years into the army of 1940. These were shallow and shapely. In 1940, however, the army adopted a new, deep design which more fully covered the ears and the back of the neck, and even most of the forehead. As soon as they began to be issued to the troops, soldiers abandoned the old name "tin hat" and gave the new name "head bucket" to the new article.

HEAD SPACE. When the barrel of a Browning heavy machine gun has been removed from the gun for cleaning or replacement, it is necessary to make the replacement with care. The proper adjustment requires a certain amount of what is called "head space" between the pieces of metal so that cartridges will seat properly and will not be ruptured by the explosion and leave the brass cylinder in the chamber to jam the next cartridge loaded. This is the "adjustment for head space" and sometimes a machine gun will not operate at all if there is "too much head space." This phrase "head space" has been applied by the army in familiar slang speech to apply to something quite different. The other meaning of head is taken, the head of a man, and "too much head space" means he acts as if his head were empty, as if he had no brains at all. It might just as well, by all the rules of logic, have been applied to a man with a "swelled head" but the American soldier hitched it onto the other meaning, the opposite meaning if you will, and there it remains.

HEDGEHOPPING. This is a mode of attack for a military plane. Coming up the gullies, sneaking behind wooded slopes, it approaches the foot troops whom it intends to bomb and machine gun, always at a low altitude. A man cannot see over a patch of trees, so the plane approaches low. It is so low that it is called "hedgehopping" just as if it had to rise a bit in its flight to clear hedges. Also, because such a trip is usually flown at relatively low altitudes, a short flight from point to point is called a "hedgehop."

HIGH BALL. The hand salute. This term is usually employed when the salute is rendered in an especially snappy manner. It is called the "high ball" by transfer from railroading slang, harking back to the days when a man beside the track waved a train on its way by signalling with a swing of his lantern on high. It is also said, in railroading, to have originated with a signal in the shape of a huge ball, drawn high on a pole, as a signal that a train might come rolling right on into the station or yards. In both cases the signal might be taken as a friendly form of greeting. Soldiers took it up and applied it to the salute. The salute itself, however, had another origin, as Colonel Youngberg of the Engineers says:

"The hand salute, which passes whenever members of the military and naval services meet, originated in the days of knighthood many centuries ago. A knight in armor, with helmet on and visor down, covering his face, could not be easily recognized and when two of them met it was customary for the stranger or junior to raise his visor. The other knight would immediately raise his visor too, which corresponded to the return of the salute. When helmets were off, no raising of visors was necessary, just as we do not salute when the head is uncovered. Knights in armor have long departed, but that custom of theirs remains in use, as the military and naval salute the world over.

"The hand salute belongs to the services. It is the recognition sign of our own kind, for all grades alike. The youngest recruit and the oldest colonel make the same salute to the general, and he makes the same return to each of them. Respect and courtesy work both ways, and all ranks and ratings stand on the common ground as sworn defenders of law and order, the honor and the safety of the state. Some people think of the salute only as a mark of subordination, but it means much more than that. As the silent password of a great body of men devoted to the service of order and right, it signifies the respect and good will necessary for proper team work, implying a sense of the common bond which unites high and low together in a worthy, unselfish purpose."

We are informed that privates in the modern German army salute one another, thus removing the idea of subordination from this method of greeting.

HIKE. To hike is for a unit to march outside of garrison or camp. This is quite a different task from such mere walking as is done by various "outing" clubs or "hiking" clubs. The soldier carries approximately fifty pounds of equipment when he hikes, including rifle, ammunition, mess kit, blanket, shelter tent, and the rest. This load on his back makes the balance quite different from that at a normal walk, and troops must be trained and hardened for hiking. Perhaps the longest hike ever made in our army was made when the Third Infantry changed station from Camp Sherman, Ohio, to Fort Snelling, Minnesota—over a thousand miles. Perhaps the most famous marches in history were the rapid trek of Sir Robert Crauford's Light Division to Talavera in 1807, sixty-two miles in less than twenty-four hours, and the bitter and hurried advance of a unit of the French Foreign Legion in 1898, across Africa from the West Coast to the sun-baked town of Fashoda on the upper reaches of the Nile. Since the infantry are those who march on foot, instead of riding like the artillery and the cavalry, this is how they are described in an army song: "I'm a fighter and a hiker in the U. S. Infantry." Brander Matthews said that "hike" was originated by the army in the Philippines, but it is believed to have been in use before that, by transposition from an older term "hike" which meant to push. A famous battalion commander of the Marine Corps got the nickname of "Hiking Hiram" because he used to ride a horse beside a column of marching troops and exhort them to "hike along heartily."

HIT THE DECK. Stolen out of the navy, of course, this phrase is now in general use in the U. S. Army. The sailor sleeps in a hammock, hung fairly high. To get him out in the morning, he is told to "hit the deck" and the words have come even into military barracks, when at first call the soldiers have to bring their feet from bunk to floor.

HIT THE SILK. In the ground forces, when skirmishers rush toward an enemy or a burst of fire from machine guns or artillery comes smashing suddenly overhead, they are said to "hit the dirt" and dive to a prone position. They do this of course to avoid danger. The aviators took up the idea of this phrase and the sailor's phrase "hit the deck" and applied it to their own circumstances. When they "bail out" and take to the parachutes they are said to "hit the silk."

HITCH. An enlistment. In civil life, when a man gets married, he is said to be "hitched" from the old phrase of hitching up a team of horses—in the days when there were enough horses around for people to know what that meant. When the soldier enlists, he is hitched to the army for three years. The word is always used in the army as a noun, never as a verb. He is "on his first hitch," and he is "finishing this hitch," and he is going to "take on another hitch."

HOLY JOE. As Mencken says, of "respectable antiquity" in the American Army, to refer to the regimental chaplain, although it is also, according to Fraser and Gibbons, used in the British Army. It is, of course, pure slang and its origin is unknown.

HOMBRE. Spanish for "man" and pronounced in the army exactly as it is pronounced in Spanish, this has become part of the regular army lingo. During frequent services for over two decades in Spanish-speaking countries like Cuba, Philippines, Mexico, and on the Border, the American soldier added this to his vocabulary, always however using it in connection with some other word: "good hombre" or "bad hombre" or "big hombre."

HOOCH. This is a word, meaning hard liquor, which the army originally gave to America. Eskimos in the north according to Mencken made a home-concocted liquor which was in their tongue called "hoochino." When American soldiers first went to Alaska to take that country over after it had been purchased from Russia in 1867, they heard the word "hoochino" and shortened it to "hooch." The army has used it ever since. It has also become part of standard Alaska vocabulary, and is widely used in certain strata throughout America.

HOP. Throughout the military service in America, a dance is a "hop." It is so at West Point. It is so at all stations. Whether the chicken or the egg came first, we cannot say. We only know this descriptive slang is universal and ancient. Be the event formal or informal, occasional or regular, from Devens to Del Rio, from Meade to McKinley, an officers' dance is a "hop"—although it should be stated that in the last fifteen years the word has been used among soldiers also when speaking of an enlisted men's dance.

HOUSEHOLD TROOPS. In the British Army this is an honored title, used to designate those units specially detailed and organized for the Royal Household. It has recently come into the American Army, even though we have no royalty. In 1940, it was decided to relieve combat troops from too much "fatigue" and routine tasks around army posts, and leave them freer than ever before for training on real soldiering. There was established in each Corps Area a "Service Command" to take care of the supply and administration of army posts, and it was said that the troops of the "Service Command" should at each army post have the duty of "housekeeping." Immediately, in soldier speech and in soldier magazines, they came to be given the slang name of "Household Troops."

HOUSEWIFE. Each squad has one, but the corporal has to carry it. It is officially listed as a "housewife" on the army tables of equipment; actually it is a small sewing kit, with needles and thread, and perhaps a button or two, for the remedying of damage to clothing when actual housewives and tailors are far away.

HOW! The army toast. The German says: *Prosit!* The Scandinavian says: *Skål!* The Frenchman says: *A votre santé!* The Britisher says: "Cheerio!" The American civilian says: "Here's looking at you!" But the American soldier says: "Here's how!" or simply "How!" One story of its origin places it among the Indians of the western plains, from the Sioux: "How kola" which was a sign of friendship, accompanied with raised hands, and says that the 7th Cavalry introduced it into the service. It was in the army before the Boxer Expedition learned the Chinese "Hao!" Another tale goes farther back, to the Seminole War regions of 1841. An Indian Chief Coacoochee is said to have noted, at a festival at Fort Cummings, that the soldiers said: "Here's luck!" or "The Old Grudge!" before drinking, asked what was the purpose and meaning of this. Gopher John, a Negro interpreter, puzzled for an explanation, said it was merely a way of saying: "How d'ye do!" The chief thereupon with great dignity raised his cup high above his head, and in a deep voice shouted: "Hough!" The incident was so impressive that officers of the 8th Infantry and the 2nd Dragoons forthwith adopted the phrase, and it spread thence through the entire army. It is characteristic and well ingrained as a military custom. One of the largest military posts in the country is the Infantry School, set among the corn liquor regions of Georgia. In days when once drinking was perhaps somewhat too free and easy on that post, a major there was confronted by a friend who passed him a brimming glass, raised another, and said: "Here's how!" To this the major replied: "I'm a graduate of the Infantry School and you don't have to show me how." Some civilians use it, but in general it is distinctively an army phrase.

HOWITZER. A short artillery piece, especially designed for high angle fire, so that it throws its shell in a steep arc, which makes it fall nearly vertically behind embankments, hills, or other forms of protection. The word, spelled "haubitzer," appears in English in the time of Marlborough and derives from German *haubitze*, which came from Bohemian. Howitzers in the United States Army are found in the Field Artillery and in the Coast Artillery. Even the Infantry in 1920 adopted the term, and called one of its units a "Howitzer Company" because it had a Stokes Mortar and a 37 mm. gun, but twenty years later abandoned that name for the more general "Cannon Company." Don't ever mistake or confuse a howitzer and a gun! It is abbreviated to "How" officially and also in speech.

HUNGRY HILL. The ancient name for the quarters of the married noncommissioned officers on a permanent army post, when these happen to be placed on a hill. It is just another name for "Soap Suds Row," and prevalent in the army for a long time but, with better pay conditions, has not been heard very much of late.

I & I. When government property in the army has "become unserviceable through fair wear and tear in the military service," it is listed by the responsible officer on an Inventory and Inspection Report, and is submitted to an official inspector who is authorized to destroy it, to order it turned in to salvage, or to order it broken up for reclamation of its component parts. If it is too bulky to be destroyed, or if it is to be salvaged and perhaps sold as junk, he marks it "I.C." From the official "I & I" paper, there has grown the verb: "to I & I." That has been transferred into slang, to mean officially condemned, and is used to refer to men and plans and plenty of other things besides property.

I.C. Meaning "Inspected and Condemned." Most frequently seen on horses, mules, heavy machinery, and on tents. Floyd Gibbons told once in the *Chicago Tribune* of going to the Mexican Border mobilization of 1916, of seeing hosts of regulars and militia under canvas so marked, and of hearing a blithe traveller say: "Wasn't it nice of the Illinois Central to lend all those tents to the army?"

I.D.R. Just the army habit of using initials applied to the book officially called Infantry Drill Regulations.

IN THE FIELD. Soldiers are either "in garrison" or "in the field." The distinction was too much for Ed Streeter's Bill, who said that, in spite of the fact that of course he wasn't in anybody's field, he could not explain it. Campaigning, maneuvering, or just camping is called being "in the field"—especially if cooking and sleeping arrangements are temporary and shelter is of a temporary character. Thus we have troops "going into the field," as contrasted to performing their daily duties on a drill ground. Thus we also have "drill regulations" and "field service regulations." The phrase "in the field" is very old in America; Washington is quoted as using it in 1779.

INCLOSURE. This is typical army. On the outside people speak of an enclosure with a letter. The army, in whatever way it encloses something, always indicates at the end of a letter the number of "inclosures"—sometimes by spelling the word out, and more usually by abbreviating it to: "Incl."

INDORSEMENT. An official and formal notation placed upon a military communication, and has nothing to do with the civilian meaning of "endorsement" as used on a check, note, or bank loan. Original letters in the army are not held by their receivers, and answered in separate letters. They are returned to the sender with whatever marks may be necessary embodied in a formal "indorsement" which is invariably numbered. Also, a letter going from a lower unit to a higher passes through "military channels" and gets an "indorsement" from each succeeding headquarters through which it passes. Thus it happens that a letter may accumulate several "indorsements," each of which is numbered in succession, and each person who receives the letter may learn the entire history of the document and its handling by the various offices through which it has passed. In civil life, an "endorsement" usually signifies "approval," but in the army an "indorsement" may just as well say: "Not favorably considered" or else, more briefly just: "Disapproved." There is a phrase in current use, common army jargon, which runs:

"By indorsement hereon," which has unpleasant connotations. Minor derelictions often cause an officer to have his fault called to his attention in an official letter, which usually "directs" that he "explain by indorsement hereon, with the return of the original communication." It is a "gig" and that's that.

INFANTRY. That branch of the army whose principal arm is the rifle, although some of its personnel are armed with pistols, carbines, machine guns, automatic rifles, hand grenades, rifle grenades, 37-mm. guns (one-pounders), and trench mortars. The following explanation of the origin of this almost childlike designation for the hardest fighting branch is taken from Robinson's *Army of the United States* published in 1848:

"It is well known that the main strength of every army long consisted in the number of knights and men-at-arms, who fought on horseback, while the common people, who fought on foot, were esteemed of little importance. The invention of gunpowder, however, worked a wonderful change in the art of war. It was a power against which Milan steel and the chain armor of the East were valueless, and enabled the foot soldiers, the people, to assert their importance. This was going backwards almost to the days of old Rome, but it was a prudent retreat, its object being to correct an abuse. The *legionarii* of Caesar's army, his foot soldiers, were far more important in his eyes than the *turmarii*, or horse. So it was with the heavy-armed men of the Athenian organization and their cavalry.

"One of the first powers of Europe to perceive this necessary change in the military organization was Spain. The wise statesmen who ruled that nation in the fifteenth century, immediately commenced the organization of a foot service; and it is not too much to say that nearly all the success of the Spanish arms depended on it. The people were called out, the famous Spanish pikemen or foot-lancers were formed, and at the head of this great army of soldiers was placed, to give it dignity, the heir apparent of the Spanish throne, the *Infante*. This arm was called the *Infanteria*, or *Infantry*."

This is a nice story. But is it true? Fuller investigation reveals that the word "infantry" originally came from the

Infantry

Italian, where *infante* or *fante* meant a servant or footman, and also a soldier serving on foot. In 1579, it is said, the "infantry" of Italy was "infamous" all over Europe. In 1612 Francis Bacon paid tribute to the fighting qualities of the doughboys of his day by saying that "the infantery" was "the nerve of an armie."

INSIGNIA. Marshal de Saxe had a recommendation on this matter: "The private soldiers are, moreover, to have a piece of brass fixed on each shoulder, with the number of the legion and the regiment upon it to which they belong, that they may at all times be easily distinguished." This is supposed to be the historical origin of regimental insignia, which in the American Army is represented by crossed weapons on the collar ornament bearing the number of the regiment. Also to be considered insignia today are the elaborate heraldic coats-of-arms or other badges, worn on the shoulder strap or lapel, to represent distinctive regimental history. In addition there is the insignia of rank, the shoulder pins of the officers, and the cloth chevrons of the noncommissioned officers.

INTERIOR GUARD. Old regulations used to make a distinction between "interior guard" and "exterior guard." The former was that posted within the camp and around the buildings. The latter was that established on distant outposts and pickets for protection against advance of the enemy. There are other phrases used to describe what used to be called "exterior guard" but the distinguishing adjective remained for years to describe "interior guard duty." But interior guard duty is not done indoors at all. It is that guard duty which is done within the limits of a military command or reservation. Details are made for twenty-four hours. Sentinels are assigned to posts of duty, three sentinels to each post, to guard particular areas or places successively. Each sentinel has to "walk post" for two hours, after which he is granted a four-hour rest before being on duty again. Each of the three groups of sentinels is called a "relief" and operates under a corporal, who is responsible that each member of his relief, or group of sentinels, knows his orders and understands his responsibilities.

ISLANDS. Always spoken of as "The Islands." We have many island possessions, and have had them all since about 1898. There are those where our troops garrison places in the Caribbean in Cuba and in Puerto Rico. There are the Hawaiian Islands. But to the American soldier there is only one place known of as "The Islands"—and that is the Philippine Islands. There the army served scattered all over the lot, in small detachments, moving much by water as well as by land, fighting insurrectos in the brush and in the open. For a full decade this service lasted, and finally implanted in the army a great tradition of "The Days of the Empire"—where the monkeys have no tails in Zamboanga, where there is a virgin in the island of Cebu, where they grow potatoes small in Iloilo, where the carabao have no hair in Mindanao. The service never speaks of the Hawaiian group as "The Islands"; and among "old soldiers" it is not "The Philippines" but always "The Islands."

JACK. Sometimes used as a synonym for "corporal" although invariably in this sense always in combination with another word, as "Acting Jack" for acting corporal or "Lance Jack" for lance corporal. It is a mere phonetic corruption as we usually hear it: "Acting Jack." Because we seldom use the term lance corporal in the American Army (where it is not now an official grade at all) the officers are likely to say:

"Acting Corporal" and the enlisted men: "Acting Jack." He is sometimes, slangily called a "Jawbone Corporal"; he delivers the goods but doesn't get the cash.

JAWBONE. A distinctively American Army term indicating credit as opposed to cash. When you get a Post Exchange credit coupon book and pay for it only at the end of the month, you are getting it "jawbone" and the day they are issued to the troops is generally called "jawbone payday." Although principally used among army people in connection with things purchased on credit to be collected for next payday on the company collection sheet, the term "jawbone" is sometimes employed also with regard to credit with down-town merchants.

The word is recorded in this use in the 1880's among cattlemen of the Northwest and is said to have been used among pioneers in Washington Territory. It is also British Army slang, said to be Canadian in origin, and used by the *London Times* in 1862. It is also said to be of very recent invention in the service, being a corruption of a phrase used by oriental merchants and hucksters in the Philippines, which means credit and has a somewhat similar sound. We have not identified that word. But when you put the American slang "bone" for dollar, and the idea of talking or "jawing" a man into letting you charge it, it is easy to see how the vivid two-syllable word must have caught on in the army. Indeed it is in some camps today used with no idea of "credit" at all but merely to label a man who talks too much. The word has been extended to mean "incomplete" or "tentative" as when the acting corporal is called a "Jawbone Corporal." Or, when in time of war temporary promotions are sometimes made, perfectly official and valid but not final and permanent, they are said to be "jawbone" promotions.

JEEP. This is one of the nicknames currently given to the tiny military motor vehicle officially termed "the ½-ton, 4 x 4, command-reconnaissance car." It is used to send out small scouting groups. It is used for an officer to dash quickly over rough ground here and there. It is used to haul light, two-wheeled antitank guns. Suddenly put into service in many parts of the army, it was widely and variously given slang names. Some of these are "midget," "kiddie car," "blitzbuggy," "bantam." But "jeep" is the term which is gaining widely and rapidly in popularity and, although perhaps not yet finally fixed, gives the best promise of any. The word is at present applied to other vehicles, and also in the Air Corps to a device known as the Link Trainer. An auto-giro is sometimes named a "jumping jeep." Indeed this word "jeep" turns up in the most amazing meanings. At one large cantonment it is very commonly used to designate a new recruit. "Jeep town" is the barracks where they are first quartered apart from other troops. "Jeep hats" and "jeep coats" and "jeep" any kind of clothing is that which is ill-fitting and no soldier wears if he can get rid of it. The application arose, of course, from the fact that initial issues of uniforms were not always well-fitting, and the awkward clothes marked the wearer as a "jeep." In this sense, its use is of course based on a figure in the popular comic strip.

JOHN. Recruit. This is the soldier's name for a recruit, and he very seldom uses the phonetically closer nickname "rookie" so popular with literary folk on the outside. A group of recruits is "a bunch of Johns." The phrase is believed by many to have originated in the practice of making models of the proper way of filling out army printed forms,—including recruiting forms—in the name of John Doe. When a recruit joins his company, a large number of new forms for his property record, pay record, equipment record, have to be filled out. Thus the constant reference to a John Doe form in the case of a new man, extends even into the organization he finally joins. There was actually a John Doe drafted into the service in June 1941, and he had a hard time convincing officials that that was really his name.

JOHN L's. Army government issue underwear, cotton as well as wool, is ankle length. By strange memories of old pictures of the Glastonbury advertisements and of the attitudes of John L. Sullivan, combined, long underwear in the army—so different from the average civilians "shorts" of this day and age—is called "John L's." This is universal with reference to the wool, but is also used in some units for the cotton too.

JUGHEAD. The army's affectionate name for the good Missouri mule, so ticketed because he is supposed to be stubborn. It has been used in a descriptive sense to speak of a "jughead battery" (a machine gun company) and "jughead artillery" (mountain artillery, which has guns which can be packed in parts), because these are often carried by mules. Kipling's "Screw Guns" were of the latter type, in general, but the American Army never uses that term. Although the army guardhouse is called the "jug" just as is the county or city jail in civil life, and that sense of "jug" is not therefore characteristically military, there is no connection at all between the guardhouse and a mule, and no connection between "jug" and "jughead."

JUMP-OFF. The beginning of a formal attack, carefully planned so that all front line units go "over the top" from an agreed "line of departure" at the same time. The word "jump-off" is thus used to designate the place from which the attack starts, and also the hour it starts. A battlefield incident may be spoken of as occurring a certain distance "beyond the jump-off" and at a certain time "after the jump-off." Early western travellers called St. Joe or any town at the end of rail or boat line, the "jumping-off place" to indicate the spot from which the covered wagon caravans started, but there has been found no connection between that and the "jump-off" which was militarized during the World War and has been military ever since.

K.P. Initials for "Kitchen Police," which means the task of cleaning up the kitchen, even when there is no kitchen. In the army to "police" is to clean up. From the very inception of company cooking, soldiers have been "detailed" in turn to do this work. In addition to cleaning the kitchen, scouring pans, washing dishes, scrubbing the mess tables and floor, soldiers on K.P. are almost invariably kept busy from reveille to retreat, peeling potatoes, slicing onions, and doing such other preparatory jobs for the cooks. The degree of pleasure which a soldier gets from this sort of work, which will make him a fine wife for some girl, is shown by the following parody on the popular "K-K-K-Katy" song of World War times:

K-K-K-K-P
Dirty old K.P.
That's the only army job that I abhor.
When the m-moon shines over the guardhouse
I'll be m-mopping up the k-k-k-kitchen floor.

In *Yip-Yip-Yaphank*, a musical review put on by the soldiers of the 1918 Camp Upton, Irving Berlin—himself one of them—appeared on the stage with scrub pail and mop and sang:

Poor little me,
I'm a K.P.;
I scrub the mess hall on bended knee.
Against my wishes
I wash the dishes
To make this world safe for democracy.

[118]

K.P.

Potato peeling, of course, is the most widely advertised act of the K.P.'s. It is said that on their first day on a wartime mobilization camp, two colored recruits were sitting on mess stools in the company kitchen more or less industriously removing the skins from potatoes and dropping the peeled spuds into the usual pan of water. "Huccum," demanded the first, "huccum dat orficer keeps callin' us K.P.?" "Hesh yo' mouf, iggorance," advised the second, "Dat am de 'breviation fo' 'Keep Peelin'.' "

KAY O. A purely phonetic, slang rendition of the initial letters of the words "Commanding Officer." Its use is very general throughout the army. The term "commanding officer" is the official way of referring to the commander of a company, battalion, or regiment, although these in oral speech are usually given as "Company Commander," "Battalion Commander," and "Regimental Commander." "Kay O," as a matter of fact, is usually applied only to a post or regimental commander, not to any of the others who by technicality seem just as well to deserve the title as the "commanding officer" of a small unit.

KHAKI. It is said that during the Indian Mutiny, a provisional force of volunteer civilians and unattached officers, formed at Mirath, was called "the Khaki Force" from the color of the uniform they adopted, "khaki" being in Persian actually "dusty." The first British regiment to wear clothing of this color is said to have been the 52nd Infantry who dyed their white uniforms a mud color before leaving for the front where they went into battle at Trimmu Ghat, on July 12, 1857. It was first used generally by British troops who went out from India for the Boer War. Roosevelt's "Rough Riders," officially the First U. S. Volunteer Cavalry in the War with Spain in 1898, introduced the khaki cotton colored cloth into our army while the regulars, as Roosevelt said, were "dressed in heavy blue woolen uniforms and catapulted into a midsummer campaign in the tropics."

For Philippine service, the regulars and the volunteers alike eventually adopted the khaki uniform, and it became generally summer wear for all U. S. troops in hot climates, including southern posts in the United States. The present color of the field uniform, both wool and cotton, is officially "olive drab"—darker with more green and less yellow than true khaki. (See o.d.) However, the cotton field uniform—whatever its exact shade—is familiarly known as "khaki" and no soldier ever speaks of his woolen uniform—whatever its shade—as "khaki." Thus a command is said to "go into khaki" in the early spring when heavy woolen uniforms, olive drab, are laid aside for the summer and all men wear cotton uniforms, also olive drab, which are universally called "khaki." A lighter shade, with more sheen, originating in the Orient is called "Hong Kong khaki" and the new regulation cottons, of a shimmering silver shade, are called "silver khaki." The word "khaki," which first meant merely a color, thus has been fixed to mean a cotton cloth.

KICK. A "kick" is a dishonorable discharge from the United States Army, given only by sentence of court-martial, and invariably in connection with a sentence to confinement at hard labor. Thus, the word is never used singly, but always with some reference to time to be served, as when a soldier, asked how his trial came out, might say: "A year and a kick."

LANCE CORPORAL. A soldier acting as corporal, with the corporal's duties, powers, and responsibilities, but without the pay and grade. Fraser and Gibbons, of the British

Army, say: "The name comes from the Italian *lancia-spezzata*, the term of the Middle Ages for a trooper who had lost his lance or horse in action, and until supplied with a fresh weapon or mount had to serve with the infantry. Horsemen being then considered superior beings, pending his reinstatement the *lancia-spezzata* ('lance' was also a synonymous word for trooper) received higher pay than his new infantry comrades, and was employed usually as assistant to the corporals of his new company. Thus the title and grade of lance corporal came in originally." It was thus it became a regular rank in the British Army one step up from the bottom. Kipling refers to it when he says the rising recruit suddenly finds "they send his name along for 'Lance.'" We have no official grade of lance corporal in the United States Army. The name is sometimes, but not very often, applied to an acting corporal, and sometimes to a private first class (who is sometimes also an acting corporal) who wears a single stripe chevron on his sleeve. He is very likely to be called a "Lance Jack."

LATRINE RUMOR. "Latrine" is the official army name for a toilet. Rumors which may be invented in and spread from idle conversation, and have no foundation, are called "Latrine Rumors." News that comes by "grapevine" is usually authentic even if not official; latrine rumors are usually false; to call any news a "latrine rumor" is to say you doubt its source.

LEAVES. In the American Army, majors wear on their shoulders gold leaves, and lieutenant colonels silver leaves. Captains look forward to the time when they will "get their leaves" and rise above the herd of company officers to command other officers rather than mainly men.

LET HER EAT. The gasoline motor feeds on liquid fuel. When the driver presses down hard on the throttle, he is said to "let her eat" and even to "pour on the coal." The word has only recently come into the army, from civilian truckers, with the influx of motors to take the place of mules.

LIAISON. A French word which, brought into our army during the World War, remained. Strictly, it means connections or contacts. It is used in the French to refer to under-cover relations between the sexes, and for mere affairs and flirtations. The French Army used it to refer to those efforts which all troops make to "keep contact" with troops to right and left. They appointed "liaison officers" to go to adjacent headquarters and report developments in that vicinity. They used it to refer to signalling contacts between ground troops and airplanes. The Americans took over all of these meanings, using the French word instead of translating it, and finally adopting it into our military language, and using it to refer to signal communications of all sorts, radio, wire telephone, signalling.

LIEUTENANT. From the French directly, from *lieu* (place) and *tenant* (holding), meaning one who holds the place of another. Except on those rare occasions when the captain is absent, the lieutenant does not *hold the place* of the captain, and the use of the word in the lower ranks is an anomaly. It is a transfer from the broader meaning of the word as "assistant" which we find in general use, as for example "political lieutenants." The British pronounce the word *lef-tenant*, and there is justification for their thus saying it, for the French with whom the military title originated, sometimes spelled it "luef" and "luftenand." In 1793 Walker gave the then current pronunciation as *lev-* or *livtenant*, but expressed the hope that "the regular sound *lewtenant* will in time become current." In unchanging England, it has not except in the navy. In America, however, we always say *lewtenant*, when the soldier does not abbreviate it to "loot" or to "looie" as he often does when speaking of a "second looie."

LIEUTENANT COLONEL. The origin of the word is, of course, from the French, and has a much more definite meaning than the mere lieutenant, for it indicates what place it is which the officer holds, the place of the colonel. The title began to be used in France and in Britain in the days of "absentee" colonels. The colonel remained a courtier at court and his "lieutenant colonel" actually commanded the regiment. James Wolfe for many years commanded the Twentieth Regiment of Foot (now the Lancashire Fusiliers) with the rank of lieutenant colonel while the real colonel of the regiment, Lord George Sackville, was not present. Such a procedure was common in England of the seventeenth and eighteenth centuries, and still is today. There absentee colonels of noble blood are colonels only in name; it has been customary for the Prince of Wales to be colonel of several regiments. It was also customary in France of the old régime, where as Boutaric says colonels purchased regiments which they often did not know and which they even retained upon promotion to brigadier. There is preserved the record of an interesting dialogue on this sort of thing between a Minister of War bent on reform and such an absentee officer:

LOUVOIS: Monsieur, your company is in very bad shape.
NOGARET: Monsieur, I did not know it.
LOUVOIS: You should know it. Have you seen it?
NOGARET: No.
LOUVOIS: You should have seen it.
NOGARET: Monsieur, I shall give the proper orders.

Louvois: You should have given them before this. You must do your part, monsieur, either announce yourself simply a courtier, or do your duty, if you are an officer.

Napoleon had no truck with absentee commanders. His colonels served with their regiments, which had no lieutenant colonels. In the American Army we have both titles, but a lieutenant colonel is second in command and assistant to the colonel. Once he was a fifth wheel with little to do unless the colonel were absent. During the World War he acted as executive officer and handled routine administration, leaving the colonel free to supervise training and fighting. More recently, the lieutenant colonelcy has become more of a rank than a job, and lieutenant colonels are used to command battalions, and for many staff assignments.

LIEUTENANT GENERAL. Until recently the grade of lieutenant general, now borne by field army commanders in the United States, was rare in our army. The title originated in the seventeenth century in England and in France, simply as a step in the list of generals, one step below the top. In Cromwell's England, the three ranks of general officers were called: "Captain-General" (the commander-in-chief, or the principal captain or leader of troops), the "Lieutenant General" (in command of the cavalry and deputy and assistant to the captain-general), and "Sergeant-Major-General" (in command of the infantry). Fraser and Gibbons say that the first and third of these were changed to "General" and "Major General." In France, according to Boutaric, the lieutenant general was the grade next below that of marshal, and for a long time was given to the commander of the army, in the field, who lost it when the campaign closed.

It became a grade only under Louis XIV, when several permanent lieutenant generals were made and given pay and privileges of the rank without the command of the army that went with it previously. The title was used in the United States, after the retirement of Grant, Sherman, and Sheridan (all generals), by a single officer who bore the title "Lieutenant General Commanding the Army," but went out again after the reorganization of 1901 with the creation of a General Staff and a Chief of Staff under the Secretary of War. It was temporarily revived in 1918, to give appropriate rank to Liggett, Bullard, and Dickman, who had been field army commanders of our forces in France. "General" Pershing was above them, and many "major generals" under them, in the corps and divisions. Again it was revived recently in 1939 for army commanders and in 1940 for commanders in certain important overseas garrisons.

LINE. This is a very common word in the army with two very common meanings and each quite different from the other. It has, of course, nothing to do with a "clothes line" nor is it anything like one, as any recruit can tell you after he has gone on a fruitless visit to the quartermaster's to draw a "skirmish line" on memorandum receipt. In strictly technical language we have the distinction between "line" and "column," between men standing abreast of one another, and men standing one behind the other. When the Top Kick says, "Fall In!" the outfit falls in in line. Troops used always to fight and advance to the attack that way, shoulder to shoulder, in nice even ranks or lines. (The "thin red line" of the British Army and all that sort of thing.) From this battlefield formation originally arose the distinction between "line" and "staff." The staff consisted of those officers who did not have to fight in the line, but whose duties kept them busy on the battlefield in other ways, as when the Baron de Marbot galloped with the Marshal's messages here and there.

From this distinction, army language slowly distinguished whole branches of the service as "line" or "staff"—the Infantry, Cavalry, Artillery were called "line" troops and the Quartermaster, Medical, and Ordnance specialists, "the staff." This distinction dates back in America at least as far as Duane's *Handbook* of 1813. In the British Army there is still another distinction—between the "guards" or royal troops, and the "line" regiments—which was imitated by colonists in America during the Revolution; they called permanent three-year troops of the Continental Army "the line" as distinguished from the temporary militia called out only temporarily while the British were threatening some particular locality. When you read in an old book about "the Pennsylvania Line" or "the Connecticut Line," the reference is to the Continentals recruited in those particular states.

LITTLE PAYDAY. A soldier is paid once a month. That is "payday." But once a month, on some other day, "canteen checks" are issued, good for credit at the Post Exchange. This is "little payday"—a modern bit of slang which means the same as "jawbone payday."

LOOT. Abbreviation for lieutenant, made by slurring the final syllables, and finally using only the first syllable. Loot, itself, not as an abbreviation but as a word in its own right, is said by Jespersen to have been introduced into the British speech from a Hindu word in the middle of the eighteenth century as a result of the wars out there. Although fully condemned by regulations, it still does exist as a practice among troops not well controlled, as instance the saying that the Americans in France in 1918 "fought the war for souvenirs" and Kipling's vivid poem entitled, "Loot."

MAE WESTS. Once used to designate a twin-turreted type of combat tank. Now general in both British and American forces for a bulging, vest-like type of emergency life-preserver, used on transports and on over-water airplanes.

MAJOR. Originally the battalion commander was called the "sergeant major" meaning a big sergeant. Thus it was used by Digges in 1579: "The Sergeant Major, by his office, is to appoint everie captayne his place." But after the sixteenth century the term was shortened to "major." Today a major is next in rank above a captain, and has for many years been considered appropriate to command a battalion. Now, in organizations expanded to war strength, a lieutenant colonel commands a battalion in the American Army and has a major as his executive.

MAKES. At West Point, when cadets are selected for promotion, the official list is said to be "the makes." Most West Point slang dies a rapid death in the army. The term "makes" drops into desuetude for a time, because all peacetime officer promotions are normally by seniority alone, until they come to pick the colonels to be promoted to brigadier general. Then the old slang is resurrected, and we hear the seniors talking again about "the makes" just as they did thirty years back. In wartime armies, most promotions are by selection again, and the word springs back into common use among all grades and ages.

MANCHU LAW. In 1912, a revolution drove the Manchus from their comfortable berths as rulers of the Chinese. In the same year, legislation in Congress provided that no officer could remain on the general staff on duty in Washington, D.C., for more than four years, and caused a great many of the apparently comfortably established "brass hats" to leave for field duty. To the American Army, that law and the regulations which put it into effect have been generally referred to as "The Manchu Law." An officer now who comes to the end of his four-year tour of duty in Washington, is still said to be "Manchu-ed."

MANEUVER. As Willoughby has pointed out in his recent book, this word has varying technical uses; sometimes it is the equivalent of movement, sometimes of plan of attack, and so on. In the old regulations, you will find drill ground evolutions spoken of as "maneuvers," but that sense has gone completely out of style. A "maneuver" is most commonly a carefully prepared field operation, undertaken for training soldiers in the field, often in large numbers and concentrated at great distances from their normal home stations. For example, Fort Riley and Gettysburg during the first decade of the twentieth century saw many "maneuvers" of regulars and guardsmen combined. More recently we have had "Army Maneuvers" for which large forces were separated into "hostile" forces and set to operate against one another with blank ammunition and campaign conditions for many consecutive days without "armistice" or artificial halt. This is what the army means when it speaks of "the maneuvers" and of "going on maneuvers."

MANUAL OF ARMS. The handling of the rifle, passing it by prescribed motions and in prescribed cadence to the various positions of "right shoulder," "present," and "port," is what the drill books describe under the title of the "manual of arms." Note that Duane in 1813 said: "For several years the most important object of military discipline was supposed to consist in the performance of certain unmeaning and frivolous motions with the firelock in the hands, which was called the Manual Exercise on that account."

MARKSMAN. This is an official title for men who qualify according to definite standards on the rifle range, and with other weapons too, but do not make the higher qualifications of "sharpshooter" and "expert." The word is very old. Without the letter s it is in Shakespeare's *Romeo and Juliet*, also in Stanyhurst's *Description of Ireland* in 1577. The town of Southampton on Long Island in 1651 complained that firearms were getting into the hands of Indians who were becoming proficient "marksmen" to the peril of the settlers. It appears in Charles James' *Universal Military Dictionary* of 1802 to describe "men expert at hitting a mark" and is in Duane in 1809. Its use as an official designation is found in British "Musketry" instructions of 1859, which prescribed that "certain of the first class shots" should be styled "marksmen." It appeared in this specific sense in America, in G. O. 86 of August 16, 1879, which declared that: "Until there shall be provided by proper legislation a system of rewards involving the expenditure of money to encourage good marksmanship, Department Commanders will grant such indulgences to the best marksmen in their commands as they may think proper."

MASTER SERGEANT. This title is purely artificial and of very recent invention. Until 1920, there were many high-ranking sergeants in the American Army, with equality of rank and pay, but a wide variety of titles, each title designating the person's particular assignment: "post commissary sergeant," "regimental sergeant major," etc. For simplicity's sake there was established the single grade of "master sergeant," the highest in the noncommissioned series, and all these various persons then held the same grade.

MATÉRIEL. The word is not misspelled. It has an e in the final syllable instead of an a. It became common in the British Army during the nineteenth century, used as an opposite to personnel. Matériel means weapons and equipment, as differentiated from supplies in general. Supply includes the furnishing of material that is consumed, like ammunition and food, but matériel—as the word is now used in the American Army—is equipment that is furnished a unit for permanent use. Ralph Waldo Emerson used the word with regard to paraphernalia around a printing shop, and it was employed by some mechanics and technicians. The American Army did not take it up very completely until the time of the World War. It sneaked into the speech of artillerymen, and has been spreading like an epidemic among folks in other branches and among staff officers who use technical phraseology. It is in our forces always accented heavily upon the last syllable.

MEAT BALL. The last day of an enlistment, when a soldier turns in his equipment, does no duty, and is discharged at 11:00 a.m. according to old custom. He gets pay and his meals for that day, and does not have to work for them. When he speaks of how much time he has to serve, he says he has "fifteen days and a meat ball." (See Roll Over)

MEAT WAGON. This is the nickname which a soldier is likely to use for the army ambulance. It ties in, in reverse perhaps, with the habit of calling the surgeon the "butcher" —even though the ambulance takes you to the surgeon. Of course it also reflects the soldier's usual deprecatory attitude toward himself. (See "dog tag," "pup tent," etc.) This attitude, common to both the land and the sea services, is excellently represented by Irving Berlin's song about the

[132]

Mechanized Force

navy: He joined the navy, to see the world, and all that he saw was the sea. The enlisted man thus assures us that he has been inveigled into a position which disappoints him. He continues: "The Atlantic isn't romantic and the Pacific isn't what it's cracked up to be." Thus you find him pretending to emphasize his disappointment with his low and disciplined station—and yet, mark this well, slyly bragging all the time that he has actually seen those seas.

MECHANIZED FORCE. After twenty years of experiment the army has finally settled down to a distinction between a motorized force and a mechanized force. The distinction has to be arbitrary, because both depend upon machines and both have motors. Our first experimental mechanized force, assembled in Maryland in 1928, was partly motorized and partly mechanized. It consisted of an assemblage of infantry moved by truck and of fighting tanks and artillery. But then the distinction was officially made: that a force is "motorized" if it moves on motors and dismounts to fight on the ground, and is "mechanized" if it fights from its vehicles. The Mechanized Cavalry brigade at Fort Knox, Kentucky, for example, had armored scout cars for reconnaissance purposes, armored combat cars (only a slightly different kind of tank), and was meant for a fast moving and quick hitting force, under the control of and used in a manner similar to, the horse cavalry of the army. In the summer of 1940 the designation "mechanized" disappeared from the army rolls: the unit had its name changed and was absorbed into an "Armored Division."

MEDAL OF HONOR. The highest decoration for bravery in the United States Army, corresponding to the Victoria Cross of the British Army. It is granted for an exceptional display of courage in action against an enemy and the performance of deeds "above and beyond the call of duty." Although the words are general, they are used only to refer to this specific decoration.

MEDICO. A member of the Medical Corps. A natural abbreviation for men accustomed to giving nicknames.

MESS. Used by the army to refer to anything which has to do with an army meal. The *New English Dictionary* notes that as "members of a party taking meals together in military life," the word goes back to 1536. It appears in America at least as early as 1760, at Fort Pitt. The mess hall is a dining hall; a mess stool is a four-legged stool of standard design, used at the mess table; a mess sergeant is one in charge of the preparation and serving of meals; a mess officer is the one who supervises such an activity; a mess kit is the individual field eating equipment of the soldier. There were days when all soldiers were trained and practised in the preparation of individual meals using these mess kits; but perfected "reserve rations" and new facilities for bringing hot food in big kettles nearer the front line have caused the old custom to be partially abandoned. Likewise, the dining room for bachelor officers is called the "Officers' Mess." When the colonel's daughter went there once for a meal, she took one look at the food on the table before her, and said: "Lord, what a mess!" Indeed the opinion has been held of company messes, of which a Chief of Staff of the Army said that, in 1917, "what the government purchased was better than what the housewife got, but what the fellow in the kitchen did to it was something awful." Things are better now.

MESS CALL. The brief and clear bugle tune to indicate the meal hour. We have for it the word of Frazier Hunt that it is the most popular call in the army. Hunt adds: "Although young training camp graduates with two silver bars on their shoulder straps may dispute it, the cold fact remains that a company's mystic *esprit de corps* is born in the mess hall and not in the captain's orderly room."

MILITIA. Originally any military force, but more recently meaning citizen soldiers as distinguished from the regulars or professionals. Adam Smith in 1776 called the militia, the citizens of military age who joined in some measure "the trade of a soldier to whatever other trade" they happened to carry on. In the United States, legally, militia is the whole body of men liable for military service, whether organized or not, whether trained or not. In 1912, regulations spoke of "the National Guard" and "the unorganized militia." As a fact, militia service is obligatory, National Guardsmen are volunteers.

MILK BATTALION. Companies of an infantry regiment are named alphabetically from A to M. The last four companies, I, K, L, and M, form the Third Battalion. It was customary, in days when battalions were accustomed to serving at separate stations and therefore to be spoken of and thought of separately, to designate this Third Battalion —by a single "lettergram" of the initial letters of its companies—as the Milk Battalion.

[137]

MILL. A soldier name for the guardhouse.

MONKEY. The "Company Monkey" is the soldier clerk, a soldier put on special duty at a typewriting machine, pencil and paper in the company orderly room. He it is who wrestles with the intricacies of balancing totals and subtotals on payday collection sheets. He it is who deciphers the awful penmanship of the captain and transforms it into neatly typed sheets, does the hunt-and-hit on the antique Remington, and turns out letters ready for signature fit to meet the discerning eye of the colonel. He is a corporal without command, but he is wise in the personal ways of the officers and their peculiarities, and soon apt at preparing papers in accordance with regulations. Gilbert Seldes tells of one:

"He found the actuality of army life comparatively pleasant, and as soon as he was made company clerk, he managed to slip in ahead of the waiting lines, so that his food had not been pawed over and the rinsing water was still clean. . . . After he had managed to get a furlough for a little Italian boy in the company, he often lay in bed while the grateful private brought him his breakfast. It was all grossly unmilitary and wrong, but Roderic liked it. . . . He worked for these men none the less. He straightened out allowance and insurance tangles, got back some nine hundred dollars for men in the company, . . . and persuaded his lieutenant to give nothing more serious than company discipline to three men who had been A.W.O.L. On the last day, when the company filed out to entrain for Camp Meade, where they would be discharged, they cheered the officers, the mess sergeant, the top kick, and the supply sergeant. Roderic, standing at the head of the company street, hating with an immortal hatred everyone of the one hundred and seventy men for whom he had worked, wondered if he could forgive them everything in return for a cheer. His question was left unanswered. The last man of the company turned beyond company headquarters. No one had waved a hand in recognition."

MONKEY DRILL. Mounted calisthenics performed as part of a soldier's training in the Cavalry, and sometimes perhaps in the horse-drawn Field Artillery. What the modern motors have done to this is a caution!

MONKEY JACKETS. During the World War the American soldier wore a tight blouse buttoned right up to a standing collar around the neck. A decade later they adopted the "roll collar" or lapel collar blouse, much more like an ordinary civilian coat. Then still a decade later, the blouse was loosened and made even more comfortable by giving it bellows slips over each shoulder blade. However, when uniforms were made in 1917 and 1918, they made them by many millions, for no one dreamed the Armistice was going to come so soon. That left a lot over. The CCC did not use them up, for the CCC was not army. They were issued sometimes to recruits, until measurements could be taken and well fitting blouses of the new pattern manufactured. They are also used in emergencies of sudden expansions. When so used, because they look tighter than the new ones, they are called "monkey jackets" by the soldiers. Many a National Guard division took some of these out of storage and for the first few weeks in camp in 1940 some of the poor devils had to wear them. Officers use it sometimes to refer to the "mess jacket" of the blue and the white uniform.

MORNING REPORT. A monthly record book in which are entered day by day the strengths (number of individuals) by grade of a unit, and also remarks giving names of the individuals involved in any changes in the figures. It is a current daily historical record, and is never destroyed. The War Department files in Washington are full of thousands of these, and they are constantly being consulted with reference to pensions, claims, burials. It is prepared by the first sergeant, initialed by the company commander, and sent to headquarters every morning. It used to record facts as from 6:00 a.m. one day to 6:00 a.m. the next; hence its name. Now it runs from midnight to midnight; but it is still called the "morning report" just the same.

MOTORIZED FORCE. A motor of course indicates an automobile. In the old days all army vehicles were drawn by horses or mules. Then in the first few years of the twentieth century, troops began using auto trucks to haul supplies. In France they called them "camions" (although this French name never stuck long in Americanese), and used them also for hauling troops long distances. But after the World War, there was a definite tendency to "motorize" certain regiments, that is to haul the supplies of infantry regiments by truck instead of by wagon. The men still marched on foot. Then they began giving the doughboys a ride, even shuttling the trucks back and forth to bring along in groups large numbers who could not be hauled on one trip of the motors. They now were getting motorized with a vengeance.

The testing of motor transportation went so far as actually to put cavalry horses on trucks to ride them long distances; the motors stopped; tail gates were lowered; animals were brought out; saddles were adjusted; troops mounted up, and the dashing cavalry was itself again. We called it "Portée cavalry"—imitating the French. In the Field Artillery they also hauled guns and caissons that way, and called it "Portée Artillery." These were motorized units, and to be carefully distinguished from mechanized units. It is proper to say that most of the artillery of the army today is motorized. Men travel in trucks. Ammunition is carried in trucks, even in the front-line batteries. The horses are gone. Special wheels and axles are put on the guns so they can be trailed along behind the trucks at high speeds. This is sometimes called truck-drawn artillery, sometimes "motorized." Most of the artillery in the army today is motorized, except the famous "show" unit of the Field Artillery at Fort Myer, with its matched horses.

M.P. See "Police."

MUD CRUSHERS. Very old army slang for the Infantry. Descriptive, like "gravel crushers," but usually derisive.

MULE SKINNERS. Soldiers who handle mules and are reputedly so "hard" they are able to skin mules alive. The phrase came in from the "outside"—but the army held to formal pack mule transportation longer than folk in civil life, so this sobriquet has lasted longer in the service than elsewhere. When the army is completely mechanized and motorized, the term—like the mule—will disappear. But in the Southwest there are still units of the cavalry division and in Panama many troops who "pack" their supplies on mules, so that "mule skinner" still remains.

MUSTER. A periodical check of the strength of units, now abolished, but once a serious formality for which mustering officers were formally appointed. Muster rolls were prepared to show all the individuals belonging to a unit, certified to by a senior who had to visit the organization and verify the actual presence of each man whose name appeared on the roll, and then forwarded to army headquarters. The practice was prescribed periodically for the regular army, and was always utilized when state militia was "mustered" into federal service, or "mustered" out. It originated in the days of ancient professional armies in which, in addition to "royal" regiments maintained in the name of the King, others were raised by indenture, with contracts specifying details as to pay, equipment, service, and allowances. Fraud entered with false figures; and Spaulding quotes an old authority to say: "Only too often did the captains 'beguile the service with lesse numbers than they are payed for.' The answer to this was adequate inspection and check; and a system of musters grew up, providing for personal verification of numbers by royal 'mustermasters.' The procedure was simple but effective, and nothing better was found down to the abolition of musters, within the present generation. 'At every musteringe or assemblinge, the captaines bill shall be called by the clarke, every man answering to his own name, marching foorth as he is called, that noe man unto twoe names make answere.' " There is ample evidence in Newhall's recent work, *Muster and Review*, that the procedure was general in the British Army in France in the fifteenth century. The bi-monthly "muster" was abolished in the American Army in 1918 and has never been reinstituted.

N.C.O. Abbreviation for Noncommissioned Officer, which includes the corporals and all the sergeants, but does not include the warrant officers and the commissioned officers from the lieutenant on up. It is used some in our army, but on the lips of the American soldier more usually you will hear "noncom's," which is of course another brief way of saying the same thing. Kipling said: "The backbone of the army is the noncommissioned man."

NATIONAL GUARD. This is sometimes erroneously called "the militia," when as a fact we have no active true militia in the United States. Volunteer units of individuals, formed into regiments under state auspices with the approval of the federal government, have come to be called the National Guard, in imitation of the National Guard formed in Paris at the time of the French Revolution, avowedly to keep the peace and to protect the government against disorderly uprisings. The first force to be so designated in the United States was a predecessor unit to the famous New York 7th Regiment, later the World War 107th Infantry, and now the 207th Antiaircraft Coast Artillery. Back in 1824 when Lafayette came to visit America, out of compliment to him as former commander of the National Guard of Paris, Companies A to F of the 2nd Field Artillery of New York were named the "Battalion of National Guards" and so called for a time until renamed. On April 14, 1869, New Jersey named its organized militia "National Guard."

NEVER MIND THE GUARD. When "Number One," i.e., that sentinel designated with that number who is always stationed at the guardhouse itself, shouts: "Turn out the guard!" it is the only time a private gives a valid order to a sergeant and a corporal. At his call, they bring the guard outside. He is required to shout that command at the approach of the commanding officer of the fort, the officer of the day, and armed parties, excepting troops at drill. If the person whose approach has thus caused the guard to be turned out does not desire to receive the formal honors or to inspect the guard, he salutes, and "Number One at the Guard House," as the sentry is called, announces to the corporal and the sergeant: "Never mind the guard!" This phrase has, in jest, come to be frequently used to call off any action.

NUMBER ONE. The origin of fantastic rumors, which are said to come most commonly from the first seat of the latrine. Another meaning is complimentary: to designate the man who stood at the head of his class at a service school, where all students are ranked numerically from top to bottom according to their academic standing.

O.D. Officer of the day. That officer detailed daily from among all officers in turn to carry out the necessary instructions for the maintenance of garrison guard. He reports to the commanding officer in person for instructions, is on duty for twenty-four hours straight, and reports to the commanding officer to be relieved by him. Since his orders always require him to make at least one round of inspection of the sentinel posts on interior guard between the hours of midnight and of reveille, he is often referred to in jest as the "officer of the night." The use of the abbreviation is very common, and acceptable, in ordinary speech. Only in the most formal and official speech is "Officer of the day" used.

o.d. Olive drab. Always used without capitals, in connection with clothing, paint, etc. This is the abbreviation which appears on official forms, as for example: "one blouse, wool, o.d.," and "o.d. paint" which is used on entrenching tools, escort wagons, permanent packing boxes, and the like. It goes directly from print into speech without change. Frazier Hunt, in *Blown in by the Draft*, tells thus of a World War recruit being called for by his parents: "They came home, and Ma and Pa and the kids all admiring him and saying how much straighter he was and how broad he looked in his new suit. 'You don't call it a suit, Ma,' George corrected, 'It's a uniform, an o.d. uniform.'" The color of the uniform is called "olive drab" because it is somewhat of the shade of pickled olives, and because it is considered drab enough not to be conspicuous, against any color background. In fact, it is a mixture, and is composed as Colonel Vere Pinter says of: "a pinch of orange, a dash of canary yellow, a drop of old rose, the same of Alice blue, a jigger of brown, and a bit of lavender." It so dates in our army from June 30, 1902.

OFFICERS' CALL. A bugle call blown at regularly stated periodical times of assembly for officers to meet at headquarters. It is also used in emergencies to call all officers to headquarters to receive immediate orders. The following words have been set to the tune:

> Young officers, old officers, field officers all.
> You're ordered to headquarters by officers' call.

Sometimes you will hear it said that the words are "really" these:

> Come to the colonel, the colonel, the colonel.
> Come to the colonel, the colonel, the colonel.

But that is an argument which cannot be settled, for all of these words to go with bugle calls are unofficial and improvised, here or there. At times special situations cause special temporary improvisations. Word for overseas movements in 1918 always came suddenly and meant quick action, so one of the regiments down on Long Island began wording their interpretation to officers' call as follows:

Min-ee-o-la! Min-ee-o-la! We go to-day!

OFFICIAL TERMS. Occasionally two members of the military establishment have a serious disagreement which cannot be ironed out by terms of friendship or administration. On the outside, someone might get fired, or resign and get a job elsewhere; but this case is not serious enough for a court-martial and you don't leave the army even for a thing like that. They do not speak, except when necessary in the performance of official duty; and they then are said to be "on official terms."

OLD ARMY. To the "old file" and the "old timer" the "old army" is always the best. After every war there have been increments of new blood into the regular service, men who came in during the war and liked the army way and remained in. The old army of the early 1900's was the army before the Spanish War, before the hosts of new lieutenants were taken in by "the crime of 1901." Nowadays, the old army is the army as it was before the World War, before the "jawbone captains" and the "boy majors" got their appointments and promotions in 1920. The army gets set in its ways and customs and tends to resent large incursions of new blood. As the sergeant said early in 1917: "It's this here war that's ruining the army, us having to take in all them civilians." The same thing was said during the expansion period of 1940 and 1941.

OLD FILE. An officer with long service and somewhat advanced in years himself. Every officer is a "file" as we have already seen, so the veteran officer is called an "old file" while the veteran soldier is called an "old timer," although occasionally also an "old file." The old files are often thought by the youngsters to be just a little queer. Once in a while you will hear one of them saying to himself: "Where are all the queer old files that were in the army when we were young lieutenants?"

OLD MAN. The commanding officer. To the officers, the "old man" is the colonel of the regiment, or the post commander. In addition, the term is often applied by soldiers in a company to their own company commander, unless they have adopted some less complimentary nickname for him. Even though a young lieutenant, he may be referred to as the "old man" if he is a company commander. Analogy is to be found in the French Army where among the soldiers and in their familiar slang the Generalissimo is referred to as "grand-père."

ONE PER CENT. The army way of saying that if I loan you a dollar until next payday you pay it back to me with one dollar extra for "interest." This is "one per cent" in soldier speech.

ONE POUNDER. A small gun which throws a shell weighing approximately one pound. It is also called a 37-mm. gun and, adopted from a French model during the World War, was used by Infantry to fire on machine gun emplacements and pill boxes, and also on enemy tanks. The type has been improved in the American service, especially in being made more adaptable for quick change of aim against rapidly moving tanks and armored cars. The pride and energy of the crew and the accuracy of the weapon may be judged from the following rhyme about it written by J. B. Howat:

Rolling along down the highway or plowing through bottomless mire,
Or up in the line where there's fighting, asking permission to fire,
Close to the plain gravel grinder where heavier guns never go,
You will find us, the one-pounder section, taking our part in the show.
In spite of the rocks and the gullies, in spite of the mud that is glue,
We'll be there in time for the action, a-lying alongside of you.
You say that we make too much racket, you say that our gun is a runt;
But it's us that you call when the bullet bursts fall, or machine guns
    turn loose on your front.
You say we attract hostile shelling, you say that our gun is a dud,
We can trim a gnat's heel with "One Round, Common Steel" and
    not draw a drop of his blood.

ORDERLY. A soldier detailed as a messenger. In various headquarters companies, certain men designated as such are permanently designated and trained to carry messages either on foot or mounted, for officers of the staff. At a post, orderlies are detailed from day to day from members of the guard and are called "Commanding Officer's Orderly."

There are also "mounted orderlies" who ride with and care for the horses of mounted officers. Also, when a distinguished personage visits an army command, an excellent soldier of long service and exceptional record is picked out and told to report to the visitor as orderly. This is a highly coveted honor, usually gained only by a man with many hash marks. In the British Army they have the title: "Orderly Officer"— which will be remembered from Kipling's: "The Orderly Officer's hokee-moot, so help him all you can." This, to them, is the officer who inspects the guard, whom we call the officer of the day. We do not use the phrase in our army. The Germans have an *Ordnanz Officer*, who is an "orderly officer" in the other sense, one who transmits messages. Similar is the French *Ordonnance Officier*. We have no Orderly Officer. There is another "orderly" in the American Army who is not an orderly at all, in any of the senses above. He is the "dining room orderly" whose duty is to care for the table ware, cut the bread, and tend to setting up the salt and pepper, ketchup and chili, and so forth. It is to be noted also that he is the only thing around the eating end of the barracks named "dining room" instead of "mess." Never a "mess orderly" by custom of the service, but always the "dining room orderly."

ORDERLY ROOM. The office of a company commander is not called an "office" nor even in common speech a "company headquarters" but is always "the orderly room" —an old phrase that has endured on the tongues of men, designating the places whence orders come.

ORDNANCE. Harvey Cushing tells how hard it is to get accustomed to army ways and speech: "Was informed yesterday that I must go ('proceed' is the word) to Governors Island for my 'ordnance'—not quite sure what that meant or how to get there, but saluted and obeyed." Rifles, guns, pistols, swords, are ordnance, of course. But the Ordnance Department of the Army at one time controlled the issue of such things as cartridge belts, haversacks, pack carriers, which—although not strictly ordnance—were so called because they were issued by the Ordnance people. The soldier goes even farther than that, and calls his "knife, fork, and spoon" ordnance in slang speech.

OUTFIT. This is the soldier word for company, troop, battery, regiment, or division. "What outfit, buddy?" was a common phrase in the A.E.F. Mencken says in his *American Language* that it is of great antiquity in the military service in the United States. It still persists: "over in our outfit," "in my last outfit," "this is a good outfit," and so on and on and on.

OUTSIDE. This is the army term for civil life, initially meaning life outside of the fort or barracks where troops were quartered. It has come to be used solely to designate civilian status: "What did you do on the outside?" and "When I get on the outside, I'll stay!" are examples of its use.

OVER THE HILL. Used always in connection with desertion, e.g., "He went over the hill," or "He is over the hill." Not mere going absent without leave, but leaving the unit without intending to return to the army. It is the army's familiar way of saying one is going to "get away from it all."

OVER THE HUMP. There is the traditional saying that the first hundred years are the hardest. In the army a soldier is said to be "over the hump" when he has completed the first half of his enlistment. This of course applies to the regular army in time of peace only, for in times of trouble all enlistments are likely to be not for any measurable period but for the "period of the emergency," and it is too difficult for easy calculation to determine when half of that is over.

OVERSEAS CAP. A close-fitting cap without a visor, capable of being readily folded and carried in the pocket when Johnny Buck is wearing his steel helmet. It was worn overseas on field service during the World War, and not permitted to be worn within the United States. It and the helmet replaced the historic "campaign hat" of the American Army. In the 1920's the air people began wearing it, on the theory that the sunshade brim of the campaign hat was unnecessary and difficult in an airplane cockpit. In the 1930's it became standard equipment for all troops in motorized units, because the wind of passage tugged too hard at the brim of the campaign hat. Its use is gradually spreading throughout the service, as more and more soldiers get mechanized or motorized. It is now officially a "garrison cap" but the name "overseas cap" persists in most army circles.

P.X. This is the soldier's way of speaking by its initials of the "Post Exchange" which he also sometimes shortens to "Post Ex." Although he still talks of "canteen checks" everywhere, he generally says "P.X." instead of "canteen"— thus demonstrating his proclivity for initials. In spite of this he says "canteen checks" instead of "P.X. checks"—probably because it is a spoken language and the latter is almost as silly and confusing to say as "Two to Duluth."

PADRE. Where has the army been, and I'll tell you what some of the origins of its speech will be. In the Philippines, troops began calling the regimental chaplain the "padre" because that was the way the Spanish-speaking natives spoke of their village priests. They still do, when they are not saying "Holy Joe" (which is more uncomplimentary). They took "padre" to France with them, and found the English Tommies already had the word, had had it for a century since Wellington's men brought it back from the Peninsular War.

PAPERWORK. This is the bane of the service, the long and never-ending "army paperwork" which is just what its abbreviated name indicates, work on paper instead of work at drill or on the rifle range or maneuver field. Army paperwork consists of answering official communications which come down asking for trivial facts when there are many important tasks ahead, of preparing payrolls, of entering issues of clothing and equipment on the proper forms properly, of keeping the morning report up to date and legible, of making out "charge sheets" for court-martial cases, or running a duty roster so that each man gets his turn at guard duty and fatigue, and so on until the poor company clerk, the top kick, and the company commander have perpetual headaches over pen and pencil instead of teaching the company how to shoot and fight. All the ideas and devices for reducing army paperwork have had no visible results toward achieving such effective brevity as Caesar's famous three-word report of a successful campaign. But there are reasons for putting things in writing as the following tale of Our Army will show: "There will always be a 'war' between the paperwork foes and the paperwork addicts. A tale has been told and retold of how a foe became an addict. The new addict issues no verbal orders these days. It seems that his conversion to paperwork took place when he told his orderly, 'Take my horse out and have him shod.'

[151]

The orderly who was a little hard of hearing or possibly didn't like horses, had the horse shot."

PARADE. A precise military formation which is actually a review of troop units and usually takes place at the hour of the firing of the sunset gun and the lowering of the flag late in the afternoon. It has no relation to the civilian "parade" which usually means a street parade—and the army carefully distinguishes between them. In the British Army a parade is any formation of troops, e.g., a drill parade or a church parade—the "parade" meaning any formation of troops, whatever the later activities or exercises of the unit. A parade is also a piece of ground; there is a parade ground where parades are held; but there is also a company parade where the company forms, and the American Army in using the term "company parade" has so far imitated the British. So far, but no farther, for "parade" by itself in the United States is a prescribed exercise or ceremony.

PARK. In the army a park is not a place of pretty trees and grass and lovers' benches. It is a place where vehicles are "parked"— if we can let language of the automobile age explain a word which was in the army long before automobiles. Thus, in the artillery, we have a gun park—the place in the open where guns and caissons are lined up for overnight, or even for more permanent stays. The word is usually used in connection with temporary camps, although it is—rarely—used in connection with more settled garrisons. If the guns are under shelter, we speak of the area as "the gun sheds" rather than as a park. The word park has also been used in connection with motor vehicles placed wheel to wheel and almost end to end in large numbers and sometimes concerning escort wagons. Thus we have "motor park" and "wagon park."

PASS. In addition to being a piece of paper, as elsewhere, a pass in the service is a status. A man is said to be "on pass" when he has permission to be absent from duty for a short period of time, with the approval of his company commander. A furlough is for a longer period and more formal. In spite of this distinction, the words are often confused in casual speech, notably when a soldier may say he is "going on pass" when actually he is getting a furlough. The pass strictly speaking is written permission for a soldier to go through the garrison gate.

PEARL DIVER. The soldier on kitchen police who washes the dishes.

PEEP. In some units the name "jeep" is applied to the larger ½ ton vehicles, not to the ¼ ton which in that case is called a "peep."

PERSHING CAP. Before the World War, the American Army garrison cap had a small flat crown and a tiny almost vertical visor. It looked puny beside the more capacious cap worn by the British, but it was regulation and prescribed for wear. General Pershing, however, as General of the Armies, had the privilege of prescribing his own uniform. He liked one with a larger, thicker crown and a longer, more horizontal vizor. This was later officially adopted for the army as a whole, and from its origin has been called the "Pershing Cap."

PIECE. The infantry soldier's rifle is not a rifle, but a "piece" in oral language; and the artilleryman's gun is also a "piece" even in official regulations for loading and firing, which describe "service of the piece" or a "field piece."

PILLBOXES. In the World War, when fighting was stabilized on the two long defensive lines reaching from the Alps to the North Sea, the machine gun dominated the battlefield. Its rapid fire prohibited advance across No Man's Land except at what appeared to be prohibitive costs. Artillery was multiplied on both sides, in an attempt by days and even weeks of preparatory fire to blast the machine guns out of place and thus permit the "poor blooming Infantry" to go forward and capture ground. To protect themselves against such preliminary bombardments, the machine gunners and their weapons were then placed in strong concrete and cement blockhouses of great thickness with only small apertures for firing. These were called "pillboxes," in the language of the time. It is interesting to recall that when the ironclad *Monitor* during the American Civil War came down the Atlantic Coast to Hampton Roads to meet the Confederate ironclad gunboat *Merrimac*, the former was called—from the appearance of her flat deck, low freeboard, and single circular turret—a "cheesebox on a raft." Both the "cheesebox" and the "pillbox" titles are indications of the human fighting man's inclination to name his fighting articles after common things in life. It is now said that the shores of England are so lined with pillboxes that they look like the shelves of a pharmacy, although this may be an exaggeration.

PILL ROLLERS. The nickname, for perfectly obvious reasons, of enlisted members of the Army's Medical Department.

PISTOL. The army "pistol" is a caliber .45 automatic weapon with vicious striking power and quick self-reloading qualities. Prior to March 29, 1911, a caliber .38 revolver had been standard equipment in the army. It is said that the

Pill Rollers

better weapon was adopted in that year principally on account of its greater shocking force, for it is capable of knocking a man down.

PITS. The protected place behind the parapet or embankment at the target end of a rifle range, where men manipulate and mark the targets during practice shooting. Usually there are no pits there at all, merely an open place between the embankment and the hillside backstop for the bullets passing over. The phrase, however, is old and comes from days when there actually were pits. It has remained in the living speech of the soldier folk long after its accuracy has passed. The army man says: "at the butts," "in the pits," "take this detail down to the pits," and "you are on pit detail."

PLATOON. Taken by the British centuries ago from the French *peloton* (a ball or group of men) in the service meaning a small body of foot soldiers. Rarely is it used in that general sense today, but usually to refer to a specific military unit, a collection of squads moved as a single unit and commanded by a lieutenant. In the Infantry, it is the fire unit in battle, the drill unit on parade ground.

POLE. Going "up the pole" and coming "down the pole" are ways of saying swearing off liquor and recommencing its use, common in civil life, but probably brought there from the army. We have spoken of "guarding the flag pole" but feel it necessary to say that officially the flag pole on an army post is never so called. It is always referred to as the "flag staff" and a flag often called "half mast" on the outside is in the army said to be "half staff."

POLICE, MILITARY. Here the word "police" is used in the civilian sense. "Military police" are specially detailed soldiers, often formed into battalions and companies, who are used for guard duty, rarely in a post or camp, most frequently in cities or towns adjacent to military posts. They are very busy on paydays. It is always an unpleasant job they have, keeping order among soldiers on pass. And they get little thanks for it from other soldiers. They are called "gold bricks," even "dog robbers," and were in bad repute in the sight-seeing A.E.F. where one of the verses of the million-verse song said:

> The M.P.'s think they won the war,
> By guarding at the café door,
> Hinky-dinky-parlez-vous.

POLICE UP. Meaning to clean up. For example, kitchen police are those who do the main scrubbing work there for the cooks, and the "Police Officer" of a garrison is not a policeman, but rather the one who is responsible for seeing that the post is kept neat and clean, stray rubbish is collected, grass is cut, etc. To do this "police work" under his supervision, he is furnished "police details" or "fatigue details" from soldiers due for such duty.

POOP. For many decades, a "poop sheet" at the Military Academy has been a list of work to be done, lesson assignments, or a schedule of activities. It remained purely local slang and West Point graduates failed to bring it effectively into the "old army." Like "skin" and "gig," however, it has strangely begun to have wide use in the mobilized army of the 1940's, in spite of the paucity of West Pointers by comparison with officers from civil life. In the Field Artillery, the word has another use, synonymous with "fire," as when a battery of guns is fired, you are said to "poop it off."

POST. Correctly speaking, a post is a place of duty. The post of the guard is at the guardhouse. The posts of officers and noncommissioned officers are their proper places in or adjacent to the ranks. When at formal parades, or at guard mount, the adjutant commands: "Posts!" each marches to his proper place. Also, a sentry post or sentinel post is the place where a sentinel is required to perform his duties, both in campaign and in peace, and "observe everything that takes place in sight or hearing." This is either a single spot, or it may be a route, around a building, or perhaps a long route covering an extended area which must be protected from fire or depredation. Thus, also, a military reservation is called a post, for it is a place of duty for the entire garrison stationed there, and circular letters from the higher headquarters are often addressed to the "Commanding Officers of all Posts and Stations." Of interest to all in the service is the opinion of the justice in the Court of Claims who said that, like the lonely lighthouse keeper, an army man is not paid a salary and given a house to live in but is rather paid a salary to live in a house on a military reservation to be there and ready for duty.

POST EXCHANGE. Originally called the "Canteen," it is the cooperative store maintained by soldiers and officers at an army post. Based on "shares" owned by company funds, selling articles at the lowest possible margin of profit consistent with avoidance of loss, and dividing profits among the companies who originally established the exchange or bought shares in it, the P.X. as it is commonly called is a true cooperative store, run for the benefit of its customers.

PRISONER CHASER. Soldiers sentenced by court-martial to short terms of confinement at hard labor, work out their punishments as "garrison prisoners." Their work is mostly out-of-doors. Guards accompany and observe them, and since regulations require these guards to walk at least eight paces in rear of their prisoners, they are called "prisoner chasers."

PRIVATE. The soldier in the army leads what appears to be a very public life. How does it come, therefore, that the great mass of them bear the official designation of "private"? He eats at a table with many others, sleeps in a squadroom with many, and even bathes in a community bath room. When he gets promoted to the higher noncommissioned grades and ceases to be a "private" his life may become more private, with often a separate room. The term has been in use in the United States Army, however, ever since the first act of Congress creating an army, September 29, 1789, specified sixty "privates" for each company of Infantry and for each company of Artillery.

In more general use, the word is found early in the modern English language to designate a man who does not hold public office. We use it thus today, with "private citizen" and "private life." Thus Higdon in 1432 remarks: "Private persones should brynge theire goodes." In the British Army it appears in 1579; "They can do no more than privat souldiors." Shakespeare used it to refer to soldiers in *Henry IV* and in *Henry V*, in the latter play using the word "private" alone without the accompanying "soldier." This may have been for purposes of metrical contraction, for it is "private soldier" in the *Abridgement of English Military Discipline* in 1686 and "private men" in the *London Gazette* in 1691. In America, Colonel Bouquet used it in 1763, Washington and John Adams later, all as "privates." But, whatever the origin, the soldier himself knows he is listed as a "private" on the official rolls, is introduced socially among his friends as "Mister," and speaks familiarly of himself not as a "private" but as "buck private."

PROVOST. This is an old military word, common to many languages. It has two separate meanings in the army, the circumstances and the use determining the sense. In accordance with its origin, it is used to identify "provost courts" which are courts set up by military forces in occupied territory for the trial of civilians for offenses against the rules of the occupying army, as we had them in the American Forces in Germany from 1919 onwards. It is also used to denote as a "provost sergeant" not a sergeant of police at all, but rather the sergeant in an army garrison in time of peace who is charged with taking care of rakes and lawn-mowers and picks and shovels used for keeping the post clean, and of assigning work to the various prisoners from the guardhouse who are doing time at "hard labor."

PULL YOUR STRIPES. A noncommissioned officer wears chevrons or stripes. He is, however, an enlisted man and mingles with other soldiers often irrespective of rank. If, when off duty, he suddenly asserts his authority, he is said to "pull his stripes." This is a recent phrase in the service, first being heard in the "new army" of the present day. It is, however, similar to an "old army" phrase long used among officers, "pull your rank," with a similar application. The officer phrase came into the service directly from West Point, where it is common cadet slang.

PUNK. Not complimentary, and totally undeserved, this is still the army name for bread, and the "punk sergeant" is the dining room orderly (or attendant) whose principal duty is cutting the bread and keeping the bread dishes properly heaped for hungry men. Punk is also used as a familiar, and not always welcome, title for the company clerk, in the army and in the Marine Corps too, although this is not as common a sobriquet for him as "company monkey."

PUP TENT. Each soldier is issued a "shelter half" as part of his full field equipment. In the field two soldiers put two halves together, prop them into the air with the aid of folding wooden rods, tiny ropes, and aluminum pegs. The result is a "shelter tent"—all of forty-three inches high at the peak and sloping to nothing at the bottom. When two soldiers crawl into one, there is a question whether the other man's elbow or the ground is the more uncomfortable. All shelter halves are made alike, with double buttons, enabling one to be used on either side; but old soldiers like to send recruits around asking for a man with a "left-hand shelter half," to fit with their "right-hand" one.

When a platoon pitches shelter tents, there seems always to be an odd man with an odd "half." He surrenders it to the first sergeant and then goes and crawls in with two complaining privates. The name "pup tent" is said to have originated in the Sixth Iowa Infantry at Memphis in 1862, when "new shelter tents were issued—a half tent to each man—which were appropriately dubbed 'pup' tents by the men." The nickname goes along, of course, with "dog" biscuit, and "dog" tags, and "dogface." For instance, we have Captain Heileman's word for it that the 15th Infantry at Chattanooga in 1865 built improvised shelters of canvas scraps and sticks and any other material handy, and called them "doghouses." In 1911 General Funston called them "dog tents."

PURPLE HEART. By an order issued at Newburgh, N. Y., August 7, 1782, General Washington prescribed the figure of a heart in purple cloth to be worn over the left breast by soldiers who should perform "any singularly meritorious action ... not only instances of unusual gallantry but also of extraordinary fidelity and essential service in any way." By General Orders No. 3, War Department, February 22, 1932, the Purple Heart was revived in the form of a medal bearing a bust of Washington and inscribed "For Military Merit." It was thus an award for acts deserving of recognition which did not, however, rate high enough to deserve the higher awards of the Distinguished Service Medal, the Distinguished Service Cross, and the Medal of Honor. It is now awarded for wounds received in action, and thus tends to replace the "wound chevron," and also the Silver Star citations given by General Pershing for conspicuous service in France. When medals are not actually worn on the uniform—for it will be noted that they are worn only on occasions of ceremony—the Purple Heart is represented by a purple ribbon.

PURSUIT PLANES. In the American Army they call the small, fast, fighting aircraft "pursuit" planes, and the designation has an interesting history. When our air service went to France in 1917-1918, it found the French naming these *chasse* planes. *Chasse* means "hunt" and is very old, as when the lords of the manor had the right of hunting over common lands and the peasants' strips too, and the peasants could not even use weapons to protect their own crops against voracious wild birds and animals. The French Army, according to Willcox, has long used the word to signify "chase" or "pursuit" with reference to action against a ground enemy. Early in the World War, planes were used almost exclusively for observation purposes, to get a birds-eye view of what was going on behind the enemy lines. Then the aviators carried pistols and fired them at one another. Then they installed machine guns. These became a special group of "fighting" planes, as distinct from the bombardment and observation planes. The purpose of their fighting was to drive away hostile observation planes. So they actually were "chase" or "pursuit" planes, and the man who took the French word *chasse*, left off an "s" and made it "chase," and then "pursuit" did not end up so badly after all. The pursuit plane is used now to drive away attacking enemy bombardment and observation planes, and it is also used with bombardment planes to drive away other planes which might approach to drive them away. The result will probably be a "dog fight" between the two sets of "pursuit" planes, but the initial purpose is clear and the name "pursuit" is still appropriate.

PUTTEE. Correctly speaking, a long strip of cloth, wound spirally from ankle to knee, to protect and support the leg. We had canvas leggings before the World War. During that fight in France, we adopted the spiral puttees which French and British had been using in the mud of the trenches. Recently, the American Army banished the "spiral puts" and returned to a canvas legging. At this last move millions of soldiers and ex-soldiers cheered. Those whose legs dwindled under the constrictions of this twisted instrument of torture were glad to see it go. *Puttee* comes from the Hindu *patti*, meaning "band" or "bandage," and was brought into the British Army from India late in the nineteenth century. Though the London *Times* in 1900 said: "The puttee leggings are excellent for peace or war, on foot or on horseback," there are many who will disagree. Among them must be counted Ring Lardner, who in writing of his four weeks in France as a war correspondent required to wear uniform remarked: "I tried on the outfit, Mr. Gibbons assisting. We managed the puttees in thirty-five minutes. It is said that a man working alone can don them in an hour, provided he is experienced." The question then arises how any soldier wearing those things ever got fully dressed to a reveille formation, unless he expended all his time in the service gaining practice, skill, and speed, or unless he had each of the two hundred and fifty members of the company helping him—all this provided he did not sleep in his breeches and puttees. Note in passing that the leather leggings worn by infantry officers, although so called, are not truly puttees at all. Nor are the canvas leggings of the more recent uniform, although often so called.

Q. Battery Q does not exist in fixed organizations of the Army, the lettered units in the various regiments not running past the letter M. So, Battery Q is of course a fictitious name. It is what the soldiers in the Field Artillery call the guardhouse. In the Infantry they sometimes call it Barracks 13.

QM. Short for "Quartermaster." In some foreign armies the Quartermaster General corresponds to what we call the Chief of Staff, that is, the principal adviser to and representative of, the general in command. Thus Lord Roberts was Quartermaster General on an Indian campaign of the British Army, and Ludendorff was Quartermaster General to Hindenburg in Germany. But in our army, ever since the days of Gates and Washington, the Quartermaster General has been the man in charge of securing and furnishing, supplies, equipment, and food. The Quartermaster Corps is a special section of the army devoted to that work. And at each post is an officer of that Corps, called the post quartermaster, who handles all supply and procurement, repair of buildings, and maintenance and operation of machinery. He is the "QM." Needless to say this clerical branch of the service comes in for a good deal of joshing and joking at the hands of the combat branches. The "quartermaster gait" is a by-word for a lazy walk. And Dolph gives the following satirical song from the days of the World War:

> Oh, we don't have to hike like the Infantry,
> Ride like the Cavalry,
> Shoot like Artillery,
> Oh, we don't have to fly over Germany,
> We are the Q.M.C.

If it rains in the early morning, but clears up in time for drill so that the boys have to go out on the line with their rifles just the same, the soldiers call it a "quartermaster rain."

QUARTERS. The sleeping accommodations of military personnel. Its uses differ. Thus, the houses occupied by commissioned officers are called "officers' quarters" and the barracks occupied by troops are called the "quarters" of the units which occupy them. A soldier prefers to use the word "barracks." "Quarters" is, also, rarely used of tent camps and improvised shelter. Although the historic phrase "quartering of troops" which appears in a Constitutional Amendment is synonymous with "billeting of troops," and is indeed older in the American language, the word "billets" has quite displaced "quarters" with regard to civilian structure taken over and temporarily occupied. Although the soldier does, indeed, prefer to use the word "barracks" rather than "quarters," the noncommissioned officer on daily detail as "Charge of Quarters" is universally so called, and "barracks" does not appear in his title at all.

Early in the Civil War occurred a famous instance of the use of the word. Grant faced his first regiment. It had just listened to flamboyant speeches by a pair of politicians and had cheered them with hurrah enthusiasm. It called on its new colonel for a speech. All he said was: "Go to your quarters!" The origin of the word "quarters" is tied to the meaning of "one-fourth." It might be a fourth of an area, of a town, region, or district. Then, broadening in use, as many words do, it acquired a looser meaning, standing for just a part or portion. As for instance when Milton said: "Scour each quarter to descry the distant foe." Then, when troops were billeted in towns, as they were for so long and so regularly before barracks were built for regular armies, the units lived in different "quarters" of the town. Hence the bugle call or drum which sounded the "call to quarters," sounded usually fifteen minutes before "Taps," as a warning that all must repair to their quarters to be in bed in time for "Taps." This is "Call to Quarters":

RAILHEAD. What sort of thing is this, called a "rail" and said to have a "head!" Like the farmer's giraffe, "There ain't no such animal!" We adopted it during the World War. Think of a railway line reaching up toward the front in France. At some point just before it comes out into the open beyond the hills or woods, it stops. No fool is going to run a locomotive up a track right in front and in view of the enemy's line, where his machine guns can spatter its front with bullets and his artillery can throw shells into its boiler. So they stop the train, and dump the supplies— ammunition for the guns or food for the men or anything else you please. This is the railhead. Farther forward these things will be carried by truck or wagon, perhaps by pack mule, or even by hand. The railhead may actually be the end of the line, or somewhat short of the end of the line; or in pleasanter or easier circumstances where the locomotive can loop around, it may not be the end of the line at all, just a station on the line. But, in soldier speech, and official tactical language too, it is called a railhead just the same. A railhead is usually heaped with piles of stuff not yet actually carried forward to the troops, with labor units for the unloading from the cars and loading onto trucks, with a few military police standing round and looking serious, and with a lot of paper requisitions, receipts, and records.

RAINMAKERS. Big guns of the heavy artillery. This title arose from a fallacious idea that the tremendous shocks of their explosions brought about rain. The idea, though false, is widespread. In 1910 the farmers of South England petitioned that the gun practice of the British fleet in the Solent might be postponed until after their crops were under cover. Sir Napier Shaw quotes official Italian experiments under Senator Blaserna who used a battery of guns to send up smoke rings to produce rain. At the end of one year, this experiment was abandoned as a waste of public money. At one time the American Congress gave $10,000 to finance the experiments of a professional rainmaker who attempted to use a similar method. Alexander McAdie relates that this was on the theory that at "a very loud noise, the raindrops would be so startled that some of them would drop. . . . General Dyneforth exploded at half-hourly intervals half a ton of rack-a-rack (dynamite), the TNT of that period, at Fort Myer. . . . It was a muggy, hot night, and space was saturated with water vapor, light showers occurred at intervals, but they occurred mostly before the explosions, and continued long after the experiments were over. As an experiment

in rainmaking, the explosions were a pronounced failure."
Then during the World War over a period of four years
meteorologists made a check of constant gunfire, and found
there was no relation between it and rain. There was so
much space in the atmosphere that the relatively small
shocks of the guns could produce no general effect. All of this
science, though, is not going to keep soldiers from calling the
big guns by the traditional title of "rainmakers." Or even
the soldiers who were in France in 1918, where it seemed
to be raining all the time, from believing that there is a
connection between the two. The thing is folklore.

RANGE. Strictly speaking, the range is the distance
from the muzzle of a weapon to the target at which it is
aimed. The squad leader announces to his men: "Range—
600" and they know how they must adjust the sights through
which they will aim. In this sense, also, is the phrase "effective range," which indicates the maximum distance at which
a rifle or gun can be effectively employed. There is another
quite different use of the word "range." You hear soldiers
say they are "on the range" or "going on the range" and
you know they are talking of the field where targets and
firing points are established for actual shooting practice
with the rifle, the place where shooting accuracy is proved by
performance, and beyond the uncertainties of barrack argument. This application is so fixed that what the civilian
calls a "shooting gallery" the soldier will call an "indoor
range."

RANK. Defined by a cynical West Pointer, overfearful
of bootlick, as: "Military standing, the *sine qua non* of privilege in the army, also the reason all second lieutenants always attend the bridge party given by the colonel's wife.

Everyone in the army ranks someone else, and is in turn ranked by someone. He who ranks you never lets you forget it." The "ranking officer" is the senior in length of service in his grade, or the senior officer in grade. "Rank has its privileges," is an old army saying. Quarters are assigned according to rank; people are supposed to be seated at dinner tables according to rank; they are arranged at courts-martial according to rank; they are promoted according to rank. It has happened that an officer coming anew to a post will select as quarters the house he likes best, provided it is occupied by an officer junior to him, and the ousting is called "ranking out." At Fort Snelling many years ago, one officer reporting for duty was senior to all officers there and so moved into the best quarters. He ousted another by so doing, and there was a move all the way down the line, as each in succession moved into the next best to the one previously occupied. Sense in such matters has begun to prevail since the World War, however, and most commanding officers will not permit "ranking out" and insist that newcomers take their choice only from vacant quarters.

RATION. Officially defined as "subsistence for one man for one day," this word is so widely understood that it would not have been included here were it not necessary to caution folk that to pronounce it to rhyme with "nation" is the mark of a civilian and a raw recruit. Old soldiers always pronounce it to rhyme with the first two syllables of "national." Another quirk of speech causes it almost always to be used only in the plural, and loosely so as to mean merely "something to eat."

RECALL. In Kingsley's *Westward Ho* (1855), we read: "The trumpets blow recall and the sailors drop back by two's and three's." Who has not seen this happen at any post in the late afternoon, when the bugles blow "Recall" (pronounced in the army with the accent on the first syllable) and the fatigue parties come in? The "Recall" is a bugle call, used to signal a retirement, as "recall from drill" or "recall from fatigue"—to mark the end of a routine period of duty. At "Fire Call" an entire garrison turns out, goes to appointed duties, and when the fire is out does not return to quarters until the commanding officer has ordered the bugler to blow "Recall." At field maneuvers and exercises, when the directors or umpires want to bring the action to an end, they have the "Recall" sounded. Half a century ago, under the old regulations, the sounding of "Recall" was an official means of ordering a retreat or retirement; but this use has disappeared from regulations, just as the louder crash of modern battle has smothered its sound on the field.

RECRUIT. A French verb *recruiter*, meaning to secure men for military service, was anglicized in the seventeenth century. It was used in the noun form in England as early as 1653. In 1709 Marlborough, in a letter to Godolphin, spoke of "recrutes" being sent to Ostend for the campaign which resulted in Malplaquet. A man so secured is thus today a "recruit" or a "rookie" until he is trained and taught to perform the duties of a soldier. The "raw recruit with the bran-new suit" has always been the butt of soldier jests.

RED LEGS. The blue uniform of the United States Army has always in modern times had long trousers. Down the outer seam of each leg runs a two-inch stripe of a color corresponding to the various branches, the arms and services. The Artillery color is red, as on the cord of the campaign hat. Hence artillerymen are called "redlegs" and the name persisted even through many years of breeches and olive drab. There was an item, however, which helped keep the name in use. Horse artillerymen wore canvas leggings faced with leather, and later high laced boots of leather. Shining this leg leather, soldiers use a red colored paste or dye polish—and how they shine it! So in many posts, the artilleryman is still a "red leg" although he does not wear the blues with the conventional stripe.

REGIMENT. Derived from words in the Latin and French indicating a rule, and thus applied to a body of troops under a single régime or rule. As used in the Low Countries, according to Barrett (1598) the word specifically meant "a number of sundry companies under the charge of a colonell." Of course, all soldiers live under the "rules and regulations for the government of the army" and the name might, logically enough, be attached to a unit of any size. The names of all military units are, however, all general in character as far as basic meaning is concerned, and they all mean a "group of men" or "group of men for battle." It is only by arbitrary designation to units of specific size, and continuance of that designation through years of usage, that we have now definite conceptions as to the character of squad, section, platoon, company, troop, battery, flight, battalion, squadron, regiment, group, brigade, division, wing, corps, and army, and of the relative position of each in the organization as a whole.

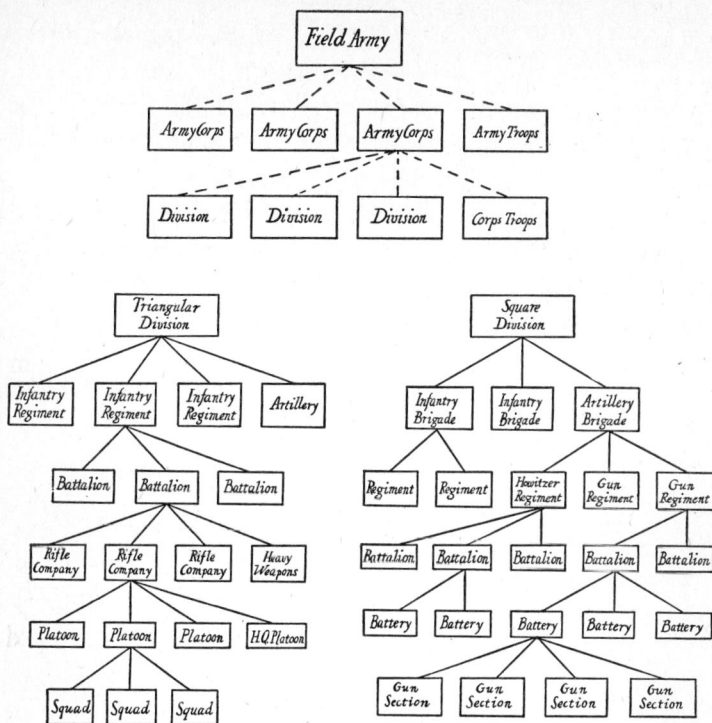

This diagram is greatly simplified. It omits many details of organization, and is designed only as an explanation of terms.

REGULARS. Professional soldiers, those who belong to the Regular Army or "standing army" as distinct from those in the militia, the National Guard, volunteers, and wartime conscripts. In the Spanish-American War they used to sing:

> You don't belong to the regulars,
> You're only a volunteer.

In church history there is a distinction between the "regular" and the "secular" clergy. The "regular" clergy were those priests who lived under the monastic rule (regle), and there is some analogy in military circles, for the Regular Army lives constantly under the special jurisdiction of court-martial, and the rules and regulations for the armed forces. In America, you will find Sir Jeffrey Amherst using the word "regulars" to Governor Hamilton of New York in 1763.

REMOUNT. A new horse in the army, and so by analogy especially in the Cavalry, often a new man in the army. The army has what is called a Remount Service, which raises or purchases new horses for the mounted units. By the habit of abbreviation these are called "remounts."

RETREAT. A bugle call, and also a formation, in the army without relation to retirement on the field of battle. The formation is held, and the call is blown, late in the afternoon just prior to the firing of the sunset gun and the lowering of the national flag from the garrison flag staff. In earlier days in the army, it was the custom for a fife and drum corps to march through the streets of a camp or garrison playing loudly, and each soldier was required to "repair to his tent or quarters" and remain therein. Nowadays they come out of quarters and stand in formation in a ceremony while the flag is being lowered. The British brought the word to America in colonial days and you will find Colonel Henry Bouquet speaking of "this evening at retreat beating" in 1760.

RE-UP. The soldier way of saying "re-enlist." The word comes from the slang method of depicting what actually happens. When enlisting and being sworn in, a man is said to "hold up his right hand" for three years. So when he does it after being discharged, he "re-ups." This is of course a peacetime phrase only, for as Kipling says: "There's no discharge in the war."

REVEILLE. An early morning bugle call of the army. Its origin is in the French *reveiller* meaning to arouse or wake from sleep. It was taken over from them by the British who first called it "revelly" in 1644, wrote it "revalley" in 1696 and in the *Perfection of Military Discipline* in 1701, and then as Fortescue points out reverted to the spelling "reveille" in the Articles of War of 1708. They have however persisted in pronouncing it, however they spelled it, as if it were still spelled "revelly" as we see in one of Kipling's *Barrack Room Ballads*, who says of a fight: "It started at Revelly an' it lasted on till dark." This bugle call is the subject of constant misinterpretation. It is not actually the one which wakes the soldier up in the morning; that blame attaches to "First Call," which is blown as a warning call fifteen minutes ahead of all army assemblies; and when "Reveille" is sounded Johnny Buck is actually out on the line in formation. In spite of all this, it is "Reveille" which has inherited all the odium which attaches to the compulsion of early rising; it was of this call that Irving Berlin wrote in 1917:

> Some day I'm going to murder the bugler
>> Some day they're going to find him dead.
> I'll amputate his reveille and step upon it heavily
>> And spend the rest of my life in bed.

The soldier, nevertheless, does not permit professional songwriters to write his songs for him, even if the professional also be a soldier. The soldier has made up his own words, which fit the tune of the reveille call itself:

> I can't get 'em up, I can't get 'em up,
>> I can't get 'em up in the morning.
> I can't get 'em up, I can't get 'em up,
>> I can't get 'em up at all.
> Corp'rals worse than the privates;
>> Sergeants worse than the corp'rals,
> Lieutenants worse than the sergeants,
>> And the Captain's worst of all.

Neither British nor Americans maintain the French pronunciation. The British pronounce it "Re-vel'ly" as the meter of the poem of Kipling well indicates; and we Americans call it "Rev'uhli."

RISE AND SHINE. In both British and American armies, a common phrase for the noncommissioned officer in charge of quarters to use when calling soldiers to get up and out for the morning reveille formation.

ROLL OVER. The last day on an enlistment when a soldier does no duty, is discharged at 11:00 a.m. and commences his first day of duty on his new enlistment the following morning. In army slang, when he speaks of how much time he still has to serve on an enlistment, he is likely to say: "Fifteen days and a roll over."

ROOKIE. A recruit, phonetically contracted. Note that this is more an officer's word than a soldier's word. It should be added that a very young recruit is in danger of being called by soldiers, a "chicken," probably from his small and tender look. But usually the men in the ranks just say "John." To call a man a recruit is to designate him as pretty green and raw, but for some reason or other to call him a "rookie" seems to imply that he is also awkward and stupid. For instance note Kipling's words in *Many Inventions*: "You can't drill, you can't walk, you can't shoot—you awful rookies."

ROSTER. A roster is a list of names. A roster of all members of each unit is mailed each month to the Adjutant General of the army. A "duty roster" is a list of names of men on which is kept a record of their availability for duty, so as to preserve an equitable division of special tasks, desirable and undesirable. The booklet in which this is kept in the company is officially termed the "duty roster" and checks for guard duty, and fatigue duty, and also for kitchen police, room orderly, and charge of quarters. The soldier who thought of "going over the hill" expressed his growl as follows:

> For it's "John, you're on the Guard List!"
> Or it's, "Buck, you're on K.P."
> Oh th' whole damn duty roster,
> There ain't no guy but me.

ROUND. A single charge of ammunition. Originally artillery projectiles (stones or "cannonballs") were round in shape, as were musket balls. Indeed there was a verbal distinction between "round shot" on the one hand and "canister," "case," "grape," and "spherical case" on the other. This, however, is not believed to be the origin of the term. The use of "round" to designate a single shot from a piece or a single cartridge for a piece grew by transposition from an entirely different meaning. Kipling's "Rounds, what rounds of a frosty night?" refers to the Orderly Officer going his "rounds" inspecting the guard. This sense is exactly that found in the *English Military Discipline* of 1686 where the fortress sentry is said to expect to have his challenge answered by "Round" or "Governor's Round" or "Sergeant Major's Round." This meaning indicated, then, going to various points in succession, or making a circuit of a position, fort, or camp. At that period, artillery pieces were not fired simultaneously as in a modern volley. They were fired in succession, around the battlements (especially in salutes) or down a line. The gunners applied this term "round" during the eighteenth century, as when the *London Gazette* in 1725 said: "The great Guns . . . fired several rounds," and Nelson in 1794: "The Garrison fired one general round." A round, then, was a single discharge from each piece.

From the idea of a "round" as a firing of the piece, the transposition was easy to the charge that was fired. "Fire one round," for instance, might be applied either way. And so a single cartridge, for rifle or artillery piece, is called a round. One of the regiments of the American Army carries as its motto a battlefield order: "Forty rounds." The use of the term in the modern sense is seen in the *Gentleman's Magazine* of 1747 which said: "Wolfe's regiment carried into the field 24 rounds a man." Our Field Artillery battery commanders today command: "One round" and each gun fires in turn at predetermined and generally understood intervals.

ROUTE ORDER. The manner of marching troops who are permitted to march out of step, to converse, smoke, and carry weapons in a manner other than that prescribed for disciplinary drills. This manner of march is adopted in the infantry at the command: "Route step!" Note that military personnel always pronounce this to rhyme with "out" and not at all the way civilian motorists do.

S.C.D. The army abbreviation for Surgeon's Certificate of Disability, upon which a soldier may be discharged as physically unfit for further service. An old tale is of a lad who enlisted too suddenly after a broken love affair, only to find it was not broken after all. He then went around picking up and looking at every piece of paper he saw and saying: "That's not it!" until the psychiatrists got at him, judged him insane, and gave him an S.C.D. Then he looked at his discharge certificate and said at last: "That's it!" The abbreviation is used to refer not merely to the surgeon's certificate itself, but also—and perhaps principally—to the discharge itself. This spreading of the meaning probably comes from the similarity of the words "Disability" and "Discharge."

S.O.S. Service of Supply. Those units in rear of a big fighting force on campaign which supply materials, munitions, and subsistence to the troops. It began to be used in our army during the World War. It was first called "Line of Communications" and then "Service of the Rear" and finally "Service of Supply." As such, it includes all installations between the docks and the front line. Since March 1942 all home ground troops except field forces have been renamed the "S.O.S." To the fighting man it is necessary, but he scorns its personnel, singing the following song:

> O, Mother take down your Service Flag,
>   Your son's in the S.O.S.
> He's S.O.L., but what the hell—
>   He's having a wonderful rest.
> He's weak and pale, but that's from ale,
>   Or else I miss my guess.
> So, Mother, take down your Service flag;
>   Your son's in the S.O.S.

SABER. You must now never call it a "sword" for in the very modern army it is a saber. It has been so in America since 1913, when the Chief of Ordnance reported that the "sword and scabbard" for enlisted men would hereafter be known as the "saber, model of 1913." The sword is a close combat weapon. Long ago even foot soldiers were loaded down with it. For example, you will find Macaulay telling that at the battle of Killecrankie in 1689 the Highlanders delivered their volleys, then "suddenly flung away their firelocks, drew their broadswords, and rushed forward." Then as the plug bayonet came in, the foot soldier discarded the sword. In 1702 the foot soldier in the British Army was encumbered with musket, bayonet, sword, in addition to powder horn, haversack and all the rest. In the course of the eighteenth century he got rid of the sword. In 1745 the 33rd Regiment left their swords in storage when they went to fight the French. In 1783 even the Royal Highlanders turned their swords in to Ordnance, and an order of July 21, 1784 abolished the sword for foot troops, except that some were still carried by noncoms. This had been the condition for some time, a Royal Warrant of 1768, specifying that sergeants should carry swords, although corporals should not.

In the Infantry today, the sword is a show piece carried only by officers at drill. It has no place in war, and is not used in wartime training periods, not even at ceremonies. The Cavalry has always thought of itself as a charging, sword-thrusting outfit. Yet in 1739, it appears, a troop was drawn up in order of battle with drawn swords, it returned swords, pulled off gloves, slung firelock, dismounted, linked horses, fired, reloaded, fixed bayonets, unfixed bayonets, unlinked horses, remounted, and drew swords. In 1765 it became the practice for dragoons to leave their swords on their horses when they fought dismounted, and in the nineteenth century, when cavalry became more and more exclusively dragoons, this became the general practice. Finally, in the American Army, the "shock action" charge of cavalry with drawn swords was abolished, and within the last decade the sword ceased to be cavalry equipment.

SALUTE. "Saluting," said Ring Lardner, "is a wear and tear on the right arm." Yet it is required by the rules of military courtesy and goes right on. It used to be that a man saluted with the hand and arm farthest from the person met, but during the World War there were so many things to teach him, that they concentrated on teaching him to salute with the right hand only.

SALVAGE. This word entered the American military vocabulary in the course of the World War. It originated officially, and in a manner quite in consonance with the historic meaning of the word at sea. To salvage was to save. Equipment dropped on the battlefield by enemy or by wounded friend was "salvaged" and brought back for disposition. There were thus piles of "salvage." And so, if a soldier found the bundle of hand grenades he was supposed to carry too heavy for his disposition, and hurled it into a nearby heap of brush, he "salvaged" it. On the other hand, if he saw something he wanted, whether on a salvage pile, or on the ground, or even actually in the possession of another outfit, he "salvaged" it for himself. It was frequently used in France; almost disappeared for two decades; and now is coming back into use again.

SAM BROWNE BELT. An officers' belt, with a leather strap over the right shoulder, presumably to help support the saber which hangs on the left side. It is a dressy article, and must be polished and have its brass fitting shined to mirror brilliance. We adopted it from the British, and strangely enough in France, where sabers were not worn, and the shoulder strap reached down the wrong side to support the pistol, which hangs on the right. It is generally said that it was invented as a field belt by a Captain Sam Browne of the British Army, who devised it to make more convenient his handling of saber and scabbard after he had lost one arm. This story is told by so good an authority as Lord Roberts, but we beg to differ. For long ages the British Army had a cross belt for officers and men of their dragoons and cavalry. There are references to them in War Office documents through the first half of the eighteenth century. From 1751 onwards however the phrases used seem to indicate single belts for the swords, and in 1795 and 1796 the reference is very plainly in the singular "shoulder belt" for the Dragoon Guards, the Heavy Cavalry, and the Light Dragoons. So it remained for three-quarters of a century. Then in 1878, according to Ffoulkes and Hopkinson, Sir Basil Montgomery of the 60th Rifles had his field belt again provided with braces for sword and pistol. In 1879, Sir James Douglas of the Royal Horse Artillery independently had the same thing done. Considering the extra weight to be carried on the belt, such coincidence is not strange. The belts were mentioned in Dress Regulations in 1894, and were officially adopted as part of the equipment on April 24, 1900. The French imitated the belt from the British, because it facilitated distant identification of the wearer as an officer. We also copied it. Its use is now optional.

SAND RAT. A soldier or officer detailed to operate, score, and care for rifle targets in the "pits" during practice, record, or competitive shooting. It is usually a hot and nasty job, hidden behind an embankment, hearing the bullets whizzing and cracking ten feet overhead, pulling target frames up and down, pasting up the holes in them with gummy water and flour.

SARGE. A familiar, phonetic abbreviation for "sergeant." It is characteristic, says Mencken, of a habit of clipping back in pronouncing words, a habit we acquired from the English and which we have applied to many other words too. The habit is not peculiar to the army, although this particular word is.

SCHOOL. The army of course is full of schools, from garrison schools in trades and occupations to a higher school for officers called the command and general staff school. Some of these are really post-graduate colleges, but only the top one, the Army War College, bears that name. School is, however, used in another sense in the service. First for the soldier comes instruction in basic drill of the individual, then training proceeds to include functions in larger and larger units. Thus it was that in the old regulations for Infantry drill we had successive chapters entitled: the school of the soldier, the school of the squad, the school of the platoon, the school of the company. The terminology has been disappearing from the books for some years, but it lives on the lips of the old soldier, and is passed on by him to new generations of recruits. The word appears in American drill books at least as early as Duane's *Handbook for Infantry* of 1813 which says: "The *Drill* answers to the word *School*, it is there the exercises are commenced and the first elements of military motion taught."

SECTION EIGHT. This is the section of a pamphlet of Army Regulations which permits the discharge ahead of his normal time of a soldier found unsuited and unadaptable for military service. It replaces one in the older volume of regulations which went by the name of Paragraph One Forty-Eight and a Half, and the outcome of the proceedings under it was a "Blue Ticket," which has already been covered.

SELECTEE. For want of anything better, this word came into use to designate the men called into the army under the Selective Training & Service Act of 1940. Some circles tried to fasten "trainee" on the troops, but they ignored it as completely as they had condemned "Sammy" in 1917. "Ducks" is good and beginning to be used in spots, but it is not as universal as "selectee," which after all is too formal and legal sounding for soldier liking, and probably will soon be replaced. "Draftee" is of course a popular term.

SENTINEL. The man on guard duty, spoken of by Anthony Munday, as early as 1584, as:

> Being his turne as he said for to watch this night,
> And breaking up sentinel when it began to be light.

SERGEANT. A name originally given to feudal tenants whose incomes were too small to be considered a knight's fee, yet who were above the common run of serfs. It had its origin in the Latin *servire* (to serve). Oman tells us that on the Continent in the later twelfth century, horsemen were classified as *miles* for the upper ranks and as *serviens* for the lower. In France it was accepted and established. In Germany the *sarient* was used not so exclusively, being mixed in with *scutifer*, *armiger*, and *strator*. These were not squires, but mounted fighting men. They therefore stood somewhere between the higher class of the fighting knighthood and the lower peasant and serf rabble. It was an honored, although an intermediate, position. From this fact the word came into modern military language for the important noncommissioned officers of the army. The sergeants, like the *serviens*, are a continuing and responsible backbone, lower in grade, than the commissioned officers. To and through the Civil War period, certainly almost universally in the eighteenth century, it was spelled "serjeant" but has for long been "sergeant" in writing, although frequently "sarge" in familiar address.

SERIAL NUMBERS. To facilitate identification and to avoid confusion of records when more than one person might have the same name, the War Department on February 8, 1918, assigned serial numbers to each enlisted man in the United States Army. These were allotted in blocks to units, and given locally to individuals. Serial Number 1 was given to Arthur B. Crean, an old soldier in the Medical Department of the army, who was a master hospital sergeant when that number was first reported for him on the muster rolls and rosters of February 1918. After the World War similar identification numbers were also assigned to commissioned officers, those of the reserves as well as those of the regulars, each preceded by the letter "O" and a dash, for example: "O—8150." The number "O—1" was given to John J. Pershing, then Chief of Staff and General of the Armies.

Also after the World War, beginning on March 1, 1919, when recruiting was resumed, a new series of identification numbers was started, beginning with the number 6,000,000. By General Order 204, on December 20, 1908, the army had established an identification tag, made of aluminum the size of a half dollar, on which were at first stamped the name, rank, and unit of the wearer. These in the British Army were called "identity discs" but in ours the soldiers promptly dubbed them "dog tags." After identification tags had gone to war and serial numbers had been provided, it was decided to eliminate from the tags the data about rank and unit, and changes in Army Regulations dated March 16, 1918, effective in June of the same year in the overseas forces, prescribed that the soldier's serial number should be stamped on his tag. This practice has been continued, and other personal data have been added.

SEVENTY-FIVE. A light field artillery piece with a bore 75 millimeters in diameter. The weapon is of French invention and is of extremely high velocity, flat trajectory, great accuracy, and exceptional rapidity of fire. It formed the major part of the barrages of the World War. It has an intricate, powerful, and very effective recoil mechanism, and is the delight of everyone who ever fires it. It is being displaced for field artillery purposes in our own army by a 105 with a range about two miles greater and a much heavier shell.

SHAM BATTLE. Although this was once a popular term, it has in the last thirty years disappeared from the army lingo, and should also disappear from civilian talk. The British used to say "sham fight" when the Americans said "sham battle." It was a field maneuver with friend and enemy firing blank cartridges at one another while the citizens stood at a safe distance and watched the racket. This was an impressive sight with plenty of noise and smoke in days when soldiers stood side by side to deliver their rifle fire. Nowadays when they are separated by many yards, such an affair would be less spectacular. It would also be less true, for there are now so many other weapons which support the front line firing line from such a great distance that they could not be seen by the people on the field and the picture would be false. The army does not go in for false pictures, even if they make a pretty show. It has therefore abandoned the old "sham battle." Nowadays they have "maneuvers" or "field exercises."

Sham Battle

SHARPSHOOTER. A rating, just above that of "marksman" and just below that of "expert," secured by a soldier who demonstrates a prescribed accuracy with the rifle by actual performance on the range. This was established as a formal qualification in the United States Army by G. O. 12, dated February 20, 1884. It has been several times said that the term "sharpshooter" originated in connection with the demonstrated accuracy of the Sharp's breechloading rifle, invented in 1848, by a man of the name of Sharp. Plausible as that may be, the explanation is obviously incorrect. The fact is that the term "sharpshooter" had been used to designate specially skilled and selected riflemen to pick off individuals of the hostile force, and so used long before the Sharp's rifle was invented. "Mounted sharpshooters" is the phrase so employed in James's *Universal Military Dictionary* in 1802, and it also appears in Duane in 1809.

SHAVETAIL. A newly commissioned second lieutenant, so called in army slang, and not very complimentary, although it is sometimes used to refer to any and every second lieutenant. It is said to have originated from the similarity in strangeness, antics, and lack of steadiness between new young officers and new young mules purchased by the government and sent to units with their tails newly and closely shaved, almost as hairless as the upper lips of a very young youngster. Prior to the World War, since which the second lieutenants have worn as insignia of rank a single gold bar, there was no shoulder insignia at all for second lieutenants. Their shoulder straps were absolutely bare. It is possible that this circumstance confirmed, even if it did not start, the idea of a second lieutenant being bare as the tail of a new mule; but we refrain from vouching for this explanation.

SHELL. A hollow metal projectile filled with explosive. Before the days of rifled artillery guns, these were round-shot. More recently they have been cylinders with a pointed nose. Loosely used, the word is employed to describe any projectile fired from an artillery gun. But, strictly speaking, there is a close distinction. "Shell" to a cannoneer means high explosive shell, and is never used by him to refer to shells containing the lead pellets called shrapnel, such always being described simply as "shrapnel," just as those containing gas or smoke are spoken of as "gas shells" or "smoke shells" respectively, never as "shell." Shrapnel, however, is not used in modern warfare.

SHORT DISCHARGE. This is not to be confused with the undesirable "bobtail" or the unsatisfactory Section Eight discharge. A "short discharge" is one which releases a man from a particular enlistment ahead of his time, but it is always in order that he may re-enlist for a special purpose, to go on a tour of foreign service, or to attend a service school, any of which might extend beyond the normal term of his current enlistment.

SHORT TIMER. This is not the opposite of "Old Timer." The one looks back, the other forward, in meaning. A short timer is one who has only a short time to serve on his current enlistment. It is also used among officers to describe one who has only a short time left to serve at a foreign station before he returns for duty in the United States.

SHOTS. Inoculation, which every soldier gets, for typhoid, tetanus, and smallpox.

SHRAPNEL. When rifled guns replaced the older smoothbore cannon, the projectiles fired were of two classes: one pointed projectile was charged to burst on impact and send its shattered metal sides to do their damage, and was called, "shell," "H.E." or "high explosive"; the other projectile timed to burst in the air above enemy troops, had a lighter case, and was filled with leaden pellets which were driven over an area and onto the enemy below, and was called "shrapnel," after its inventor, Major Henry Shrapnel of the Royal Artillery. Shrapnel itself is now out of favor in modern armies, high explosive shells being more effective although the word is generally used for "shell fragments."

SICK CALL. After breakfast and before the hour for the commencement of drill, a time is appointed for all soldiers who feel they need medical attention to report to the military hospital or dispensary. This itself is "sick call" and soldiers speak of "going on sick call." A noncommissioned officer writes their names down in the "sick book" or "sick report," has the list signed by the captain, and takes them to have their complaints heard or their ills administered to. The bugle music which announces this time each morning is "Sick Call" too.

SIDE ARMS. The artilleryman wears a pistol, for self-protection against nearby foes. The officers and the noncom leaders have pistols for self-defense, for their main business is leading men, not shooting—for which the rifle is best at all distances except for close-in self-defense. These pistols, worn at the belt, are called "side arms." But the rifleman has a bayonet for close combat, also worn at the side of the belt and also included in the term "side arms"—as Kipling uses it in his poem called "Belts" in which he tells how an Irish regiment and English cavalry had a famous fracas in Dublin. Likewise, the sabers worn at ceremonies and at garrison drill by officers and by high-ranking noncoms, are also called "side arms." Thus it is quite common to hear it said that members of a court-martial wear "side arms" or that members of the guard never remove their "side arms" even when not on post as sentries. At mess, you often hear soldiers ask for "the side arms" and this is something different again. It means sugar for the coffee or for the cereal, and this meaning comes perhaps from the little jutting handles of the heavy government issue sugar bowl, or perhaps from the fact that the coffee is the main thing (like the rifle) and the sugar just something extra (like the bayonet). The term "side arms" is also sometimes used to indicate salt and pepper on the mess table.

SILVER STARS. On the ribbon of the Victory Medal issued to all members of the army who were in service during the World War, there are attached special horizontal bars, or clasps, to denote the various engagements in which the wearer participated. When the ribbon is worn without the medal, the number of bars is represented by an equal number of bronze stars affixed to the ribbon. Among these bronze stars, you will often find one or more of silver. The silver star represents a "citation" for gallantry in action. It means that the man was mentioned in formal orders praising his performance. In England it is "mentioned in despatches." Sometimes the orders giving a "citation" are from the War Department, sometimes from the headquarters of a force in the field commanded by a general officer. But now a separate Silver Star Medal is replacing the silver stars on the Victory Medal ribbon.

SKIPPER. Originating in the navy, where the captain of any vessel is in command of the ship, the term "skipper" has become quite general in the United States in referring to any officer of the grade of captain, and sometimes even to lieutenants who happen to command companies. A captain may be thus addressed by his lieutenants without undue sense of familiarity, and still without that excessive impression of formality which a constant use of the title "captain" might create.

SLACKS. A word which we adopted from the British during the World War, to refer to long trousers, wool or cotton, worn as part of the uniform. The long white and blue trousers worn with the white uniform in the tropics and with the blues in the states, are never so called. The name arose by contrast with the tight fitting breeches with which soldiers' legs had been bound for many decades while at work. When first authorized after the War they were off-duty clothes only. A person relaxed from the tight constrictions of drill ground uniform, and wore "slacks." Now that long trousers are regulation for wear when on duty, and are even worn with leggings in the field, it is possible that the name "slacks" may die out, for there is no contrast, as there was none with the whites and the blues. The word, however, tends to remain, because it is briefer than "trousers." Then, too, trousers are worn sometimes clear, and then on field service wrapped within canvas leggings. So the tendency is growing to speak of these occasions respectively as "wearing slacks" and as "wearing leggings."

SLUM. The permanent army slang name for a meat stew which is convenient to make and to keep hot in large quantities, and is therefore a common item in army menus. It is usually a conglomeration of chopped beef, onions, potatoes, and gravy. John Kieran of the *New York Times* peered suspiciously at the pot of stew, "slum," or mulligan which was boiling away on the field kitchen and asked: "What's in this?" To which the mess sergeant retorted: "Why, have you lost anything?" At any rate, "slum" is characteristic of the army. At West Point a popular football song exhorts the team as the "Sons of slum and gravy." One of the host of army parodies improvised during the World War by the soldiers overseas, ran:

> The meat was rotten and the spuds were bum,
> They cooked them together and they called it slum.

SLUM BURNER. Since slum is supposed to be the typical army food, a slum burner is often any soldier who eats army food, just as a horse or a mule is a hayburner. Slum burner is also used to describe an army cook, and also a rolling kitchen.

SLUM CANNON. The rolling kitchen used during the World War was a huge stove mounted on a single axle and two wheels. It had a jointed and hinged smokestack which, when folded down in a horizontal position, somewhat resembled an eight-inch cannon on a caisson. Since slum was common food for those up front—when they got any— this rolling kitchen was soon nicknamed a "slum cannon" or "slum gun" or "soup cannon." That type of field kitchen is now going out of use, and the term is being changed to "slum wagon."

SLUM DIVER. Any soldier, that is of course any person who feeds on slum.

SLUSHING OIL. In the artillery there is a light oil used to slush out the barrel of a gun after it has been fired. The task is superficial and in no way represents a thorough job of cleaning. As far as cleaning is concerned, it is little more than a lick and a promise. So when a man in barracks is talking loud and long and with little value or effect, some artilleryman is likely to tell him to stop slinging slushing oil.

SNEAKING AND PEEPING. The popular slang phrase for that portion of basic military training called officially "scouting and patrolling."

SOAP SUDS ROW. An old name for the married non-commissioned officers' quarters in a permanent army garrison, when not on a hill. Vorheis Richeson in *Our Army* tells us that the name comes "from the army days when every married soldier's wife took in officers' washing to help keep up the house on the pitiful pay of an enlisted man." Times are better now. The phrase might have been taken with a smile twenty-five years back, now it is accepted more as an insult.

SOLDIER'S MEDAL. Most medals for deeds of heroism are given for acts performed in the face of the enemy, or at least in time of war. Often, however, some soldier does a courageous deed in circumstances under which civilians would be rewarded with the Carnegie Life Saving Medal or the Red Cross Rescue Medal. Some of these cases happened, however, to be rescues from drowning at sea or in navigable waters of the United States, so that the heroes could be given the Treasury Life Saving Medal, originally instituted for the old Revenue Cutter Service or the Coast Guard. The War Department eventually considered that it should have a suitable award of its own for cases of this sort; so at its suggestion Congress by Act of July 3, 1926, authorized the Soldier's Medal "For Valor" for men who might distinguish themselves by "heroism not involving actual conflict with the enemy." It has been given for saving persons from drowning, for rescuing an unconscious woman from a burning building, for disarming a maniac flourishing and firing a loaded rifle, for removing dangerous explosives from a burning area at an ammunition depot fire. When not actually being worn, it is represented by a blue ribbon with thirteen alternate stripes of red and white.

SOUND OFF. At formal parade and at guard mount, there is a moment when the adjutant commands: "Sound off!" The band, or the field music, will then march the length of the line and back to its original position, playing. From this situation, when loud music comes from one group while all the others merely stand and listen, comes the army slang term "sound off"—meaning to talk loud enough and persistently enough to drown out other conversation.

SPECIAL COURT. Sometimes in speech abbreviated to "a special," this is itself short for "special court-martial"—an official term of particular meaning. There is nothing so special about it. It is merely distinguished from other courts-martial because it is legally defined as a court-martial that may be appointed by a regimental commander and that consists of "not less than three" officers, usually five. The soldier was applying a general sense of the word "special" when he wrote to his colonel, inaccurately of course, as follows: "I know that I will be court-martialed, but I know you will want to see justice done to a soldier of yours and I hope that you will see that I get a fair trial. I just wanted to write you to see if I couldn't get one of them special court-martials in my case."

[196]

SPECIAL ORDER. An official order from a military headquarters containing specific directions to separate individuals, which primarily concern only those individuals. Special orders are not published to the entire command, as general orders are; copies are furnished only those concerned. Special orders usually deal with matter directing travel, change of duties or assignments of individuals, appointments as members of courts-martial or boards. Inasmuch as appointments of staff officers, who may speak in the name of the commander at some later time, are considered to be of interest to the entire command, they are usually announced in general orders, and thus form an exception to the usual rule.

SPECIAL ORDERS OF A SENTINEL. Those separate instructions given the different sentinels of the guard. The special orders for a sentinel post always start with an exact description of the route the sentinel must march. They include the warnings and directions peculiar to that particular post, telling of what places and things on that post the sentinel must be especially watchful. They are, therefore, a supplement to the "General Orders of a Sentinel" which he must also learn.

SPIN. This originated in the Air Corps, has got into flying everywhere, and even into non-flying circles. It is common to say a man is "in a spin" or "in a tail spin" to indicate that he is all confused in his mind.

SPRINGFIELD. Officially described as the "U. S. Magazine Rifle Calibre .30 Model 1903," this weapon when not called a "piece" is familiarly called the Springfield, because originally manufactured at the Springfield Armory. There was an earlier Springfield in use before that, a single-shot weapon made in various models in 1870, 1873, and 1876. It was very popular in the "old army" but was supplanted by the Krag-Jorgenson repeating rifle just before the Spanish War. Then in 1903, the new Springfield took the palm, and has proved to be a shooting weapon of an accuracy not surpassed by any other military rifle before or since. It is now in the process of being supplanted by an automatic reloader, called the Garand from its inventor, which is easier on the rifleman who has to fire rapidly. The Springfield was first issued to the cadets at West Point in the spring of 1904, and issue to the entire army was completed in December of the same year.

SQUAD. A squad is another of those words with a specific historic meaning that bears no relation to its present use. It is said to have been adopted by the British from the French, whose *escouade* meant a small number of men, but is also said to represent a "square" of men, how big a square is not stated. In 1757 Washington said: "Divide your men into as many squads as there are sergeants," but in modern days the sergeant has commanded a "section" and the corporal a "squad." The "section" has disappeared from some infantry organizations, but we still have the "squad." In the period when troops formed in double rank, a squad was eight men. Now the squad may vary in size between six to twelve men, including the corporal and the sergeant. In the barracks where men sleep, you will find the place is called a "squad-room" even though it might be occupied by several squads, as most of them are large enough to be.

SQUADRON. The word is from the Italian, where it originally meant a small force drawn up in square formation. It later was used to designate a relatively small unit, as when Digges in 1579 spoke of "every squadron or bodie of the watche." In 1617 a Britisher spoke of "foot in two squadrons of 250 each." In 1656 it was being gradually "most commonly appropriated to Horsmen." Thus we read in Joel Barlow's *Columbiad* of 1809, of "battalioned infantry and squadroned horse." The clear distinction is however quite recent. In 1890 a Secretary of War was speaking of "battalions" of cavalry, although in the same decade G. O. 36 (War Department) of 1899 specifies "three squadrons of four troops each." In the American Army today, the regiment of Infantry has its battalions and companies, and the regiment of Cavalry its squadrons and troops. To attempt to define the exact date of the shift of phrase amidst overlapping usage is quite beyond us here, for as Villiers says in *The Rehearsal*: "To have a long relation of Squadrons here, and Squadrons there: what is that but a dull prolixity?" First organized and named during the World War, there is an aviation unit known as a "squadron," of a size between a "flight" and a "group."

SQUADS EAST AND WEST. In the drill regulations recently discarded in favor of the new "simplified drill," a large proportion of the movements were by squads: "Squads Right," "Squads Left," "Right by Squads," "Left by Squads," "On Right into Line" and "On Left into Line" executed successively by squads. From this circumstance the soldier called his daily stint of "close order drill" by the slang phrase: "Squads East and West." Although squad movements and squad commands like "Squads Right" have disappeared from the drill books, the phrase "Squads East and West" is still frequently heard to refer to drill.

SQUARE DIVISION. The army division is a definite fighting unit, supported by troops permanently assigned to see to its food and ammunition. It is a self-contained unit, able to take care of itself. Since 1917 the American Division had contained one brigade of Artillery and two brigades of Infantry. The two brigades of Infantry had two regiments of Infantry as basic attack and groundholding troops. They gave the pattern in their battle formation to the division, two in front and two in support. Hence the name "square" division. All National Guard divisions long retained this form. The name is new, although the organization is old. The name was invented to distinguish this type from the new type of the "triangular" division. (See diagram, p. 173)

STABLE CALL. In the army, a barn is never a barn. That word may do well enough for farmers, but not for soldiers. It is always "the stables" of a unit—and note that it is in the plural. And "Stable Call" in the Cavalry and other units that have horses (and there still are a few) is that bugle call which announces the time for cleaning stables and grooming horses and mules.

STABLE GUARD. When all the garrison is asleep, there is a guard detailed to protect against theft of valuable articles and destruction by fire. But ordinary guardsmen do not suffice in the stables. So there is also detailed, sometimes out of every troop, a "stable guard" not to march in the darkness of the night "in a military manner" but rather to give skilled care to horses." If a horse is down and cannot get up himself, the stable guard is there to guide him. If a horse gets his forelegs caught over a halter rope, the stable guard is there to help him out.

STAFF. In the army there is always the "line" and the "staff" and constant bantering between the two, which mounts sometimes to open criticism of the "brass hats" by the fighting troops. There is the "technical and administrative" staff consisting of the noncombatant troops and their chiefs, the Medical Department and the surgeon, the adjutants and the Adjutant General's Department, the quartermasters and the Quartermaster Corps, the Judge Advocate General's Department, and the Ordnance Department, also the Inspector General's Department, Finance Department, Chemical Warfare Service, and Morale Branch. There is also the "General Staff"—the big one topside in Washington. General Staff Corps officers are also detailed with troops to assist a general officer in arranging the details of orders and applications of policy after the general has made his decision. In his book on the A.E.F., General Liggett has said:

"Officers and men of the line dragging themselves down a road after days and weeks in hell, fighting day and night, lousy, muddy, unutterably tired, emotionally shaken, their comrades dead or broken, now and then encountered a spick-and-span staff officer sizing them up with a presumably critical eye.

" 'Who the hell is he?' the line asked profanely of one another, and 'What's the large idea? I'd think a lot more of you, you swivel-chair warrior, if you were as dirty as I am and had been where I've been.'

"When the line is not bitter about the staff, it is inclined to be humorously acid, and, it is asking a little too much, I suspect, of human nature in the mass, ever to hope that it will change. Even the old soldier who knows exactly how valuable the staff is, or the line soldier who has seen staff duty himself, is not always able to choke down his instinctive resentment at the contrast in the relative safety and comfort of the staff.

"Yet it is just as illogical of a line soldier to hoot at the staff as it would be for a railroad fireman to mutter at the superintendent of motive power or at the chemists who test the thermal units in his coal and the lime content of the water used in locomotive boilers. Without a staff an army could not peel a potato, let alone fire a shot. In all its branches the staff is nothing more than an instrument for the service of the line, and the more mechanical and complex war grows the greater importance the staff must take on. It is the nervous system and the brain center of the army."

A staff or truncheon or baton was anciently carried as a token of command. Officers not assigned to regiments or battalions for duty were then said to be "attached to the staff" and more recently to be "on the staff" or even to be the staff itself. This according to Duane's compilation of over a century ago.

STAND. You "stand" a formation when you are present at it. The reason is of course that practically every formation in the army was originally a matter of standing in line. Thus the soldier is said to "stand reveille," "stand retreat," "stand inspection." The phrase is old in the army. It quickly becomes common speech. It is universal in the new forces.

STANDING ARMY. This is a traditional English word to refer to a professional force kept permanently in service and in pay, or "on foot" as they say in England, instead of an army raised for special occasions and then disbanded, as were the English armies before the seventeenth century. It is our "regular army." The phrase nearly got into the Constitution of the United States, for some of the Convention delegates wanted to prevent the nation from having one and almost succeeded. It is seldom used in America, and usually in the service only in jest, when soldiers have to stand in ranks overlong.

STARS. The insignia of a general officer. A brigadier wears one on each shoulder, a major general two, a lieutenant general—when we have one, which has not been always—three, and a "full" general four. The word "full" general in the army is exactly paralleled by a similar phrase in academic life, where they speak of a "full" professor to distinguish that grade from the grades of "assistant professor" and "associate professor," who are socially called professors. And there is an exact analogy in speaking of a "full colonel" to distinguish him from a lieutenant colonel. But, back to the stars! You will hear it said that a man is "looking for a star," meaning that he is working his hardest and doing everything possible to get promotion from colonel to brigadier.

STARS AND STRIPES. The general public may think that these are the red, white, and blue of the national colors. The soldier uses the name to designate baked beans; and often, after they have been served at mess, you will hear a soldier softly whistle the tune heard after the sunset gun has just been fired: "Oh, say can you see?" In Elizabethan days, writers could say openly what they mean, but we cannot. So, if this is not immediately comprehensible, I may go on to help you understand by saying that French soldier slang refers to beans as "soixante-quinze"—the 75 mm. artillery gun with a sharp explosion.

STATION. Not in the army a police station, which is called the guardhouse, or a railway station, but just a synonym for the word "post." There are military stations which are not posts, and neither camps nor forts. These are referred to as "stations" and the phrase appears at the top of many a circular letter sent out to all "posts, camps, and stations." A huge supply depot, a manufacturing arsenal, a storage warehouse, a flying field—these are all covered by the word "station."

STOOGE. The company clerk is also frequently called the "company stooge" for he is the butt of many jests and buffer for many blows.

STRAIT JACKET. This is the soldier's familiar name for the World War type of blouse, left over in large quantities and used for issue to recruits, as has already been explained under "monkey jacket." It is not only that this straight-collared blouse is tight-fitting, but that it is now usually seen only on recruits who do not yet know their way about.

STRIKER. A soldier who does extra-duty work for an officer for extra pay. The British call him a "batman." The soldiers themselves call him a "dog robber." Such work may consist of personal service like blackening boots, polishing brass, etc., or may consist of caring for a furnace, keeping the lawn cut, acting as a house servant to clean floors and wash dishes, or it may consist in caring for the officer's horse, whether private mount or government animal assigned to the officer. Sometimes he is called an "orderly."

STRIPES. The soldier rarely calls things by their official names. The corporal's or sergeant's chevrons are referred to as his "stripes," and wound chevrons are "wound stripes."

SUB-CALIBER PAYDAY. Those occasions in the old army when soldiers, contrary to regulations, took advantage of days of clothing issue to carry outside articles furnished them for use in the military service and sell them to downtown secondhand clothes dealers. (It is a court-martial offense, for technically all such clothing belongs still to the government.) It was then transferred to the day when Post Exchange checks were issued. The word has been slowly going out of use, although its corollary "jawbone payday" continues strong. It was called sub-caliber because it was subordinate to a real payday, and because, in the Coast Artillery especially, often instead of firing the really big guns for much of their target practice, they fired smaller ones from the same site and called it "sub-caliber target practice."

SUICIDE SQUAD. This is the soldier way of referring to any machine gun outfit. When machine guns were first introduced into the American Army they were manned by volunteers. Because it was found that these weapons became the immediate target for hostile artillery, the men who operated them were called "the suicide squad." The word was so apt to the occasion that it has remained in use, even after the size of the machine gun unit had risen from a small platoon, to a company per regiment, to a company per battalion, indeed even to entire battalions of machine guns.

SUMMARY COURT. The army even abbreviates such important proceedings as trials for infraction of law, and uses this shortened form to speak of a summary court-martial. It is not as summary as it sounds. In fact there is actually no such thing as a "drum head court-martial" in the service. A "summary court" consists of one officer, usually a field officer, who tries simple cases. There is no judge advocate to prosecute for the "government" and—although regulations permit—there is usually no defense counsel. Witnesses are called, including any that the accused soldier desires brought. In spite of the simplified form of the judicial process before a summary court-martial, the usual protections for the accused are present. Further, the "summary court" cannot award any punishment in excess of confinement at hard labor for one month and loss of two thirds pay for a like period.

SUNSHINER. Service in the tropics is especially hard on people from the temperate zones, until they get acclimated to the hot weather in steady doses. When they have reached that stage, they are likely to get the nickname "sunshiner" and the word carries with it an implication that the constant tropical heat has made them a bit queer in the head. The word is frequently used in the army, on account of the army's tropical service in Panama, the Philippines, Puerto Rico, and Cuba, although strictly speaking it is a civilian phrase which has been adopted in the tropics by the army and brought back to the states by the army for use in occasionally referring to an "old timer" who is bound to complain of the cold, even that mild form of cold which he meets in Virginia.

Since army rules keep people on duty in tropical climates only two years, everyone on a "tour" out there is expecting to take a transport home. Transports arrive and leave for the States at regular intervals, and soon each man looks forward to the time when he too will be sailing on one. If the tropical climate begins to affect him, it is common army talk to say he has "missed too many boats." The phrase is occasionally used by civilians in such countries too, but the army has adopted it as its own, although it is of use only in the tropics of course. On the contrary, the American Army has taken into the Caribbean and East Asiatic tropics the typically American phrase "squaw man" and has applied it to a soldier who lives with or marries a native.

SUPER-NUMERARY PAYDAY. Those occasions when credit is extended for the securing of books of canteen checks, moving picture theater tickets, etc. Just another name for a "jawbone payday" and because it is long winded and rather exact—not nearly so popular a way of saying the same thing. The soldier always prefers the short slang way.

TAKE ON. To enlist. Usually used with regard to a man's re-enlistment, as for example: "I'm going to take on for three more years" or "I'm going to take on another stack." The "stack" is, of course, by analogy to the stack of chips at a poker table, after the first stack is finished.

TANK. During the first two years of the World War, the British devised what is now known as the "tank" to provide an armored vehicle capable of advance in the face of machine gun fire which checked ordinary foot troops. Developed under the name of "Heavy Section, Machine Gun Corps," these huge vehicles were tested and brought to the front in great secrecy. To keep the secret and make more surprising their first use on the actual front in France, they were transported across England, over the Channel, and through France covered with heavy tarpaulin screens. But what were these huge things? The question was bound to be asked. So they were spoken of by a commonplace name. "Cistern" and "reservoir" were suggested, and rejected. "Tank" was brief. It was adopted for the time being, and "tanks" they have remained to this day. Those which were used in the Cavalry regiments of America's recent armored force were on account of a legal technicality officially termed "combat cars"—and so called by the cavalrymen. But, unofficially and to the army at large and the public, they are all "tanks." It is worth noting that no widely accepted nickname has ever been devised. Recently they have occasionally been dubbed "galloping G. I. cans" but that is so much longer than the true title that the betting is it will not become universal.

TAPS. A bugle call of peculiar beauty and solemnity. It is the last call blown by buglers at night, and is always sounded at military funerals. In its present form, the call was devised during the Civil War by General Daniel Butterfield, to replace the earlier "Lights Out" or "Tattoo" which had been inaugurated at West Point in 1840 and used in some regiments during the Mexican War in connection with funerals. It did not become general for this use until the Army of the Potomac gave it popularity during the Peninsular Campaign in 1862, where it is said first to have been used by Battery A of the 3rd Artillery. The following account of the composition of the current music is given by J. G. Mantle on the authority of Oliver W. Norton, who was Butterfield's brigade bugler:

"General Butterfield was especially interested in the invention of bugle calls. In the month of July 1862, the Army of the Potomac rested in camp at Harrison's Landing, a point on the James River in Virginia. It was immediately after the seven days of fighting before Richmond. The losses had been heavy and the army was recruiting to strength after the long struggle. Day and night the long winding valley and the hills on either side echoed to the bugle calls that marked the rhythm of camp life.

"The old order for 'Lights Out' which had been inherited from the earliest West Point memories, sounded discordant and unsuitable to the sensitive musical ear of General Butterfield. He immediately began turning over in his mind such musical phrases as seemed to convey the suggestion of the peace and quiet of the camp—of rest after labor, a sense of pause after the activities of the day. Having settled upon a combination of notes that seemed to him to be in tune with the sentiment of a sleeping camp of soldiers, he summoned his bugler, Norton, and began to teach him the new call, whistling the notes over many times and correcting their time and phrasing. At last satisfied with the result, he jotted notes down with a pencil on the back of an envelope.

[208]

"That same night Butterfield's old brigade was the first to hear the lingering refrain of the new call, and the next morning buglers of other camps nearby—for its music had carried far among the hills—began to inquire as to its meaning, and to ask permission to learn it. Wherever it was heard, it arrested immediate attention and lingered in the memory. It passed from army corps to army corps with great rapidity and was finally substituted, by general orders, for the old 'lights out' call, and printed in army regulations. Its use in the military burial service, both by the veterans of the war and by the United States regular army, has added greatly to the tenderness of its associations. There are few musical phrases in the world held in greater reverence, and the sounding of 'Taps' when the day's work is done will hush the noisiest and most boisterous throng."

As for other bugle calls in the army, we have words written to fit the notes and sentiment of "Taps," as follows:

Fades the light
And afar
Goeth Day
Cometh night;
And a star
Leadeth all,
Speedeth all,
To their rest.

Love, good night,
Must thou go
When the day
And the night

Leave me so?
Fare thee well;
Day is done,
Night is on.

When your last
Day is past,
From afar
Some bright star
O'er your grave
Watch will keep
While you sleep
With the brave.

This is one of the few topics in army life to which the soldier has not applied his characteristically sardonic and devastating habit of familiar interpretation. Yet, as with the music of "Reveille" the soldier is not satisfied to let others have their way with army things, and he has made up his own, somewhat simpler version of words to fit "Taps"—simpler, but still dignified, a version sung by the 2nd Massachusetts in the Civil War.

> Put out the lights.
> Put out the lights.
> Out the lights, out the lights.
> Put out the *lights*—and go to bed.
> And go—to bed—

It is very easy for an outsider to get sentimental over some phases of soldier life, but the soldier doesn't get sentimental. But if anything can make him so, it is the music of taps. A soldier of the World War wrote home of it to Ithaca in these words:

" 'Call to Quarters' blows, and soon 'Taps' will lay us to rest for this day. As surely as the bugle calls of the day (save 'Mess Call' and 'Pay Call') are to be damned, those of the night are to be blessed. Particularly 'Taps.' No matter how a man wearies of this army, here is one call he wouldn't mind hearing every night his life through. It seems to us something more than beautiful music. In a way it symbolizes and humanizes this army that rides your neck all day, whispering at night that, after all, the army wishes you well, and that it's all for the good of the service. There are men who, if they go to bed before it sounds, lie awake and await it, much as the devout await Benediction. The grind, the disgust, the oath, the spur—these it obliterates, saying all our prayers for us and sending us quietly to sleep, better ready for another day."

TATTOO. This bugle call was originally designated the Taptoe from its significance as a drum beat warning to close all taverns and canteens at military garrisons, and to soldiers that they must repair to their barracks or quarters for the night. Before the invention of "Taps" during the Civil War, this was the "lights out" call. Now "Taps" is blown at 11:00 p.m. and all must be in bed at that time; "Tattoo" is sounded at 9:00 p.m. and is a signal that lights in sleeping quarters must be out; those in hallways and day rooms may remain lighted until "Taps." The ancient "Taptoe" was done by drums, with a detail making a circuit of the garrison. An old drill book of the British Army of 1701, quoted by Fraser and Gibbons, says: "The Tattoo or Taptoe: used in Garrison or upon the Rounds to warn both the Soldiers and the Inhabitants when they ought to repair to their Quarters or their Guard, for when the Taptoe is returned to the body of the Guard a Warning piece ought to be shot off, after which no person ought to be out of his Quarters or from his Post unless the Watch-word be given him." In America, Duane's *Handbook for Infantry* of 1813 says that "the sergeants and corporals call the roll at Taptoe time." This attendance check continued into the twentieth century, shifted to the hour of "Taps"; in fact it is still made in many regiments, but it is made as a "bed check" and not by hauling men out into the night and onto line.

THREE CHEERS. At parade and at guard mount, when the adjutant gives the command, "Sound off!" it is customary for the band to play three short chords or flourishes before starting the march music. These are called the "Three Cheers."

THROW THE BOOK AT HIM. To give a man being tried by court-martial the maximum sentence allowable according to the "Manual for Courts-Martial," which is the army's canon of military law. A court may be lenient if it chooses; but if it imposes the maximum, it is said to "throw the book at him."

TIN HAT. The trench helmet used during the World War, which has been retained as combat headgear for the army on active campaign. A change in shape is causing a change in the slang name, to "head bucket."

TOMMY-TOT. For the last five years, reserve officer graduates of colleges with R.O.T.C. units have been permitted to apply for a year of active duty in the regular army to the number of 1,000, of whom fifty may be selected for permanent commissions in the regulars at the end of a year. Because the legislation which made this procedure possible was sponsored by Congressman Thomason, and is called the Thomason Act, the temporary appointees at many army posts have been dubbed: "Tommy-Tots."

TOP KICK. This is the top sergeant—the first sergeant of a company, troop, or battery, so called because he is the highest ranking sergeant in the company and the direct right-hand man of the company commander. If the company commander doesn't think he should be, there is something wrong with the company commander. Many a top sergeant gives a grumpy or recalcitrant soldier "what-for" without saying a word higher up, and he keeps many worries from the shoulders of the "old man." One said to a group of recruits once: "I can talk to you that way, but the officers can't. That's what a sergeant's in the army for." He is also called just "The Top."

TOP-SIDE. Pidgin English, taken from a mode of language common to Chinamen and Filipinos, and meaning upstairs or a higher level. For instance, on the island of Corregidor, the garrison groups on the sloping hill are, according to their locations, called "Top-Side," "Middle-Side," and "Bottom-Side." The word has been brought from the Far East with folk returning from foreign service tours, and is occasionally heard at garrisons at home. It is also frequently used to speak of persons in authority, or at a higher headquarters, as "Those Top-Side."

TOUR. A detail of duty for a specified period. An officer has "a tour of foreign service" and that does not mean a round-the-world sightseeing trip. It means he goes to an overseas garrison and stays there for two years before he comes back to some home station. There is a "tour" of four years as a limit on certain desirable stations like Washington, D. C., and for undesirable stations along the Mexican border. There is a "tour" of staff duty which is by law limited to four years, at the end of which the officer must return to duty with troops. There is a "tour of guard" when officers or individuals are assigned to guard duty continuously for several weeks or months at a time. Outside of the army, to "tour" means to travel; in the army it means to "stay." An officer who goes to the service school at Fort Leavenworth as a student is there only temporarily; one who remains there from year to year as an instructor or with the troops has a "tour" at Leavenworth.

TRAVEL ORDER. An official order emanating from competent high authority, usually from the War Department except in cases where authority has been specially delegated to issue such orders, which requires military personnel to go from one place to another distant post of duty, usually involving travel by train or transport. Strict government auditors require that every such order shall contain such a phrase as: "The travel directed is necessary in the military service."

TRIANGULAR DIVISION. Less than a decade old in the American Army, first used descriptively and now as an official term. Infantry is the basic divisional fighting force; other troops in a division exist only to support it. Unlike the "square" division, the new "triangular" division has no brigade organization at all. It contains three Infantry regiments, two to be in the advance, and one in support, according to modern tactical theories, and hence its name. The square division of the World War numbered nearly 30,000 men, that of the post-war period 18,500. The new "triangular" division counts 15,000, making up for its smaller numbers with greater mobility and a larger percentage of automatic rapid-fire weapons. Its triangular character is emphasized by the fact that its artillery does not merely support the Infantry in mass, but has been broken into separate battalions each of which goes with one of the three Infantry regiments. All infantry divisions are triangular. (See diagram, p. 173)

TROOP. Originally the word "troop" was very general in meaning, indicating "a body of soldiers" from the French *troupe*. But by the time of Marlborough it had come to mean horsemen, and he said: "The troops might embark with the two regiments of foot," thus making a clear distinction. The captain of Cavalry commands a "troop," which is the administrative and tactical unit corresponding to the Infantry company and the Artillery battery. In the American Army, it was called a company of cavalry until 1883, when the change in title was officially made, by War Department orders. The word "troop" has certain grammatical peculiarities of usage. In the singular it has a specific meaning, referring to a Cavalry unit of a certain character. In the plural it may have two meanings. As Jespersen has pointed out, Macaulay in one passage speaks of "two troops of rebel horse" and shortly thereafter of forces consisting of "two thousand five hundred regular troops," using the collective plural as a synonym of "soldiers."

TROOPER. In spite of the various meanings of the word "troop," a trooper is exclusively a Cavalry soldier.

Trooper

TWO-STRIPER. A corporal wears two-striped chevrons. He is in the army therefore a "two-striper." The navy has a similar phrase for a lieutenant. The army captain has two bars or stripes perhaps on his shoulders, but the name is not applied to him.

TWO-THIRDS. "Six and two-thirds" and "Three and two-thirds" are court-martial sentences as army speech has them. Officially such sentences are: "To be confined at hard labor at such place as the reviewing authority may direct for six months and to forfeit two-thirds of your pay per month for a like period"—and similarly for three months. But that is too long for military men to say, especially if they can find a shorter way of saying it. Court-martial sentences are limited by law to fines not in excess of two-thirds of a man's pay for any given period. So the fraction is standardized, and easily handled in the phrases above. Indeed, now that finance officers have arranged so that courts will assess fines in dollars and cents, the amount is still the "two-thirds" and though the written sentence may speak of the dollars and cents, conversation about the sentence always reverts to the easy and traditional "two-thirds."

TYPEWRITER. A popular name for the machine gun, so called because its somewhat irregular and still rhythmic sound resembles that of a typewriter in skilled and rapid hands. The word came into the American Army during the World War from the British. The French soldier slang for a machine gun was *moulin à café* or "coffee grinder."

VICTORY MEDAL. A service medal awarded all soldiers who were in the armies of the Allied powers in the World War, accompanied by bars to indicate the campaigns in which the wearer participated. The color of the ribbon is that of a rainbow, to show the large variety of nations that were allied in the struggle against the Central Powers.

VOLUNTEER. A citizen who freely offers himself for enlistment. They do it all the time in the Regulars and the National Guard in time of peace, but the word is usually used only with respect to volunteers in time of war, to distinguish them from those who are "called" as militiamen or "drafted" in a nation-wide conscription. The volunteer has always had a monopoly of the hurrah's, as the following version of an old song, made at the time of the Revolution, will indicate:

> Here's to the squire who goes on parade,
> Here's to the citizen soldier.
> Here's to the merchant who fights for his trade,
> Whom danger increasing makes bolder.
>   Let mirth appear,
>   Every heart cheer,
> The toast that I give is the brave volunteer.

WALKIE TALKIE. A portable field radio. For many years they have had portable field radios in the American Army, to be moved by car or truck, but this one is so small that it is packed right along on a soldier's back, so he can walk and talk by radio telephone at the same time. It is therefore, it would seem inevitably, a "walkie talkie."

WARD MAN. A Medical Department soldier who works in a hospital ward, and so by extension any Medical Department soldier at all.

WINDJAMMER. Nothing to do with the sea or sailing ships, the term "windjammer" is a favorite epithet soldiers give to a bugler.

WOODBUTCHER. Just the soldier's customary uncomplimentary and absolutely inoffensive way of speaking of the "artificer" of the old days, "company mechanic" of the new days, who does much odd carpentry work around the barracks, along with his repairing of equipment and weapons.

YARD BIRD. The exact origin of this has eluded discovery so far. A "yard bird" in the modern drafted army is a recruit, perhaps because he is confined as a quarantine measure for a time. The word gets extended to apply to any soldier who draws only $21 per month, which is all the new man gets for the first three months. It is believed to have been transposed from the Military Academy at West Point where an "area bird" is a cadet so ill adapted to military matters that he is constantly checked up and is compelled for punishment to walk a number of hours in the quadrangle of the cadet barracks there. The phrase is common at all Air Corps training fields.

YEARLING. The word comes of course from the race tracks and in America has been widely applied to young horses. It was adopted at West Point as a label for what in other colleges are called sophomores. The implication is, naturally, that the individual has completed one year. With the induction in 1940 of citizens for (as it appeared then) one year of military training, the word found a new use. "Selectee" is too complicated to come glibly off the tongue. "Draftee" was frowned upon. "Yearling" was brief and abrupt, and indicated one who was in for only one year.

YELLOW LEGS. Cavalrymen. Like the "Red Legs" nickname for artillerymen, this derives from the color of the stripe down the trouser legs of the blue uniform in the cavalry branch of the service.

ZOMBIE. This word is generally heard "on the outside" to refer to a person a little queer in the head or short on brains. It is believed to have been employed first in the United States in a moving picture representing life in Haiti to indicate a person brought to life by various voodoo practices. Then a rum drink appeared in cocktail bars, given that name, because it was supposed to make a person a "zombie" who had to be led around and told what to do. In the army of today, it is acquiring popularity to refer to a man who falls in the lowest grouping on the official General Classification Test given to all incoming soldiers.

NEW TERMS AND TERMS NOT IN
GENERAL USE

In spite of the author's determination to include in the body of this work only those words apparently well fixed and generally in use, it has been considered desirable to append the following "Jim Crow" list of certain others not yet fully or widely accepted.

A.A. Official abbreviation for antiaircraft, not widely employed in oral speech.

ACE. During first World War used for an aviator who had downed five enemy planes. Beginning to come back again.

A.F. Short for audio-frequency in the Signal Corps.

ALBATROSS. Chicken for dinner.

ANGEL'S WHISPER. This is a general term for army bugle calls which seems to be gaining ground in the air service.

ARM DROPPER. An artillery section chief who signals by dropping his raised arm for the gun to fire.

ARMY BANJO. A shovel.

ARMY STRAWBERRIES. Prunes on the mess table.

A.T. Official abbreviation for antitank, not widely employed orally.

AWKWARD SQUAD. A group of recruits undergoing instruction, an older phrase used much more on the outside than in the army.

B-ACHE. West Point slang for a complaint or to complain which lasts hardly as long as the second lieutenant's gold bars.

BAIL OUT. To jump by parachute from an airplane. Common in civil as in military aviation, the phrase is not solely of the service.

BATHTUB. A motorcycle sidecar.

BATTERY ACID. A derogatory and infrequently used word for coffee, known principally in motorized and mechanized units.

B.C. In the Field Artillery, this is official for "Battery Commander" and is being used familiarly in place of "The Old Man" or "The Captain."

BEAM, FLYING THE IRON. Following a railroad, used among aviators only.

BEAM, FLYING THE WET. Following a river, among aviators only.

BEAN GUN. Sometimes heard for "slum cannon"—never popular and going out of use.

BEND THE THROTTLE. Go at high speed by plane or motor vehicle.

BENZINE BOARD. Another name for B-Board, which lights a fire under a man and runs him out of the service.

BIG BERTHAS. Large artillery guns, much used in the last war, now almost dead. Sometimes applied to women of avoirdupois.

BISCUIT GUN. Among aviators, principally at flying schools, it is an imaginary weapon to shoot food to a student pilot who cannot land.

BLACK HAWK. The black army necktie, as distinguished from the khaki necktie. Of recent growth and not yet fixed.

BLISTERFOOT. A new word for infantryman.

BLITZ. British for aerial bombardment, not yet widely adopted here.

BLITZ IT. To polish—from the trade name of a brass polishing cloth.

BLITZ WAGON. Temporarily used for "blitz buggy" and "jeep" but now almost dead.

BLITZES. Air patrols. Not frequent.

BLOOD. Ketchup. Not so widely used as "redeye" which is the common civilian term taken into the army.

BOUDOIR. Sometimes used for a squad tent.

BRASS. Empty cartridge shells, really an abbreviation, not slang.

BRASS POUNDER. In the Signal Corps, as in civil life, one who operates a telegraph key. Not typically army.

BROWN. To curry favor with superiors or to "bootlick."

BUCK SERGEANT. By analogy with "buck private" (just a big male) this is used for "plain sergeant" to distinguish the "sergeant" from the "staff sergeant" and the "first sergeant" and the "technical sergeant" and the "sergeant major" and "master sergeant." The phrase is new since 1940, and was never heard in the old army, where even a duty sergeant with only three stripes had a great deal of prestige.

BUDDY SEAT. Sidecar of a motorcycle.

BUNKER. German word for the steel casemates set into the ground, typical of the Siegfried and Maginot Lines, separated from larger fortifications. Not native American but adopted by our newspapers from Europe, and has not taken hold here, where our only bunkers on land are on golf courses.

BUNKIE. A personal friend or special "buddy"—adopted by the army from general civilian use.

BUTTON CHOPPER. Occasionally used to describe a laundry.

BUTTON UP. To close up a tank, used in armored units.

BUTTONED UP. Used but seldom in America, and then to mean that orders have been carried out; in Britain it means "all prepared."

[223]

BY THE SEAT OF YOUR PANTS. Flying an airplane by personal skill and judgment rather than by instrument.

CADRE. The skilled nucleus of a training unit, which receives, and trains an increment of recruits.

CANS. Headphones of radio operators.

CANTEEN SOLDIER. Occasionally heard to describe a soldier wearing non-regulation clothing or insignia, presumably bought at the canteen.

CARE BOY. Driver of a tank.

CARRIER PIGEON. An officer's messenger.

CHINA CLIPPER. A recent synonym for "bubble dancer."

CHINESE LANDING. A plane landing One Wing Low.

CLUTCH, SLIPPING THE. In motor units, this is beginning to replace "working the bolt" and "slushing oil" of other branches.

COCKPIT FOG. Mental confusion, in the air corps only.

COMB. To make a detailed inspection. Used in the armored forces concerning tank inspections.

COMBAT TEAM. Perhaps too purely official, though widely used. This is a "team" of an infantry regiment with its habitually supporting battalion.

COMMISSARIES. Groceries sold by the commissary officer. Army for many years, now pretty well understood elsewhere.

CPX. "Command Post Exercise," which is a maneuver without troops except headquarters staffs and communications personnel.

C.Q. This is unofficial but common written and oral abbreviation for the term "charge of quarters" meaning the non-commissioned officer in charge of quarters.

CRAWL. To admonish. Pure West Point. Dies quickly in service.

CROSS-BAR HOTEL. In a few places used to name the guardhouse.

DAWN PATROLLING. To arise before "Reveille." Frequent in Air Corps units, rare elsewhere.

DEVIL'S PIANO. A phrase for a machine gun, not very popular.

DING HOW. Chinese for "very good" but not much used except by men who themselves have served in the Far East.

DODO. A flying cadet who has not yet been up solo.

DOGFACE. Civilian slang which the soldier applies to himself.

DOG SHOW. Foot inspection.

DOODLE BUG. A tank.

DOWNHILL. The last part of an enlistment, when a man is well "over the hump."

DRIVE UP. For "come here" this is in the modern motor army beginning to replace the old "front and center."

D.S. Official abbreviation for "detached service"—rarely used except officially.

DUFFLE BAG. Navy for "barracks bag" but also occasionally heard in the army.

EGG IN YOUR BEER. Too much of a good thing.

FLAK. German for antiaircraft fire, widely used by British airmen, and by Americans in Britain.

FLASH GUN. A machine gun recently devised to spot aimed "hits" on targets by electric spotlight. Useful in training. The phrase is growing in popularity as the device is in use.

FLOWER POT. Turret of an airplane.

FLYING CHINESE. See Chinese Landing.

FLY BY THE SEAT OF YOUR PANTS. Rule-of-thumb, unscientific flying.

FOUR-BY-FOUR. A four-wheeled car with four-wheel drive. Technical.

FROG STICKER. Sometimes used for "bayonet" although "pigsticker" is more common, and neither is in general use.

GALVANIZED GELDING. A tank has been so named by some bright boy whom few have followed.

GARRISON SHOES. Calfskin dress shoes, rarely used except officially.

GAS HOUSE. A beer garden or saloon.

GAS HOUSE GANG. Chemical warfare instructors, who take units for training through a "gas house" to expose them "masked" to war gases.

GASOLINE COWBOY. Driver of a tank.

GEAR. In the Signal Corps this is used for radio equipment, introduced probably from the Navy and Marines who call all a man's personal equipment "gear."

GOAT. The junior officer of a regiment, that is, the newest second lieutenant. The word is used almost exclusively among officers, and rarely by any but a West Point graduate.

GOOF BURNER. One who uses marihuana. Rarely heard.

GOO GOOS. Filipinos, but used only locally in the Islands.

GOON. A rare bit with the same meaning as "zombie."

GO-TO-HELL CAPS. A recent local invention for "overseas" caps, not popular. Also used for the "fatigue hat."

GRANDMA. Sometimes employed for low gear on motor vehicles.

GRASSHOPPER. Small, low-speed, low-flying plane for artillery observation.

[226]

GREASE MONKEY. In the Air Corps and in Armored Forces, used for a mechanic's assistant; from civil life, not distinctively army.

GROUND LOOP. Aviator's bad landing applied to mental confusion.

HANGAR PILOT. A man who talks his flying on the ground.

HARD ROLLED. An occasional army name for "tailor made" cigarettes.

HASH-BURNER. Recorded, but never yet heard by me, for "cook."

HELL BUGGY. Just one of those other names for a tank which were not adopted at large.

HOOK. Recent slang invention in the new army, the term "get the hook" means get a hypodermic injection of anti-typhoid or anti-tetanus serum.

HOT CRATE. A high-speed airplane, only rarely heard. "Crate" alone is common.

HYPO HAPPY. An amateur photographer.

IRON HORSES. Tanks occasionally.

IRON PONIES. Limited in use, to refer to motorcycles.

KANGAROO. Recently invented for the sergeant of the guard.

KIWI. An Air Corps officer who does not fly a plane. Taken from the British, who got the title from a ground bird of Australia. Rarely used outside the Air Corps, and not widely there.

KNUCKLE BUSTER. The army mechanic's new term for a crescent wrench.

LATRINE-OGRAM. Of limited use as a substitute for "latrine rumor."

LAUNDRY. Slang at a flying field for a board of officers which passes on cadets' qualifications and "washes out" many of them.

LAY AN EGG. In the Air Corps, to drop a bomb.

LEADEN BREECHES. Lazy, but of scant popularity.

LIMP LINE. Truly descriptive slang for the men reporting at Sick Call in the morning; not widely used as yet, but should be.

LITTLE POISON. A new nickname for the one-pounder which is as yet not well established.

LIVE AMMUNITION. As distinct from dummy ammunition or blank cartridges, this is fully loaded with powder and ball. The phrase is general and fixed, but rarely used except when talking shop.

MARFAK. Sometimes heard for butter, not widely.

MESS GEAR. A soldier's individual mess equipment.

MOTOR POOL. At some stations the automobile garage is called the "pool" because all vehicles are "pooled" for common use.

NAPPY. The company barber.

NETTED. A Signal Corps term to indicate sender and receiver are tuned in to one another. Outside that Corps it is not used; inside it is used pretty freely.

NINETY-DAY WONDER. Used in the last war to designate graduates of the three months' officers' training camps; now being revived for the products of the Officers' Candidate Schools.

OFF THE BEAM. Incorrect. Used thus freely in the Air Corps, applying a technical flying phrase.

ON THE CARPET. Sometimes used to mean called before the commanding officer for admonition or punishment, taken over of course from civil speech.

ON THE DOUBLE. Not a formal word of command, but frequent in oral orders to a man told to do something in a hurry, as to come or go at "double time," which is the army version of a run.

PARATROOPS. Illegitimate abbreviation for parachute troops, rapidly growing in use, probably because of the awkwardness of the adjacent "t's" in the proper phrase.

PFC. Private First Class.

P.I.D. Widely used a few years ago for "Proposed Infantry Division," which was the forerunner of our present triangular divisions.

PONTON. The standard dictionaries have finally admitted that constant use by the army of "ponton" for "pontoon" made the word acceptable "for military use." Some of our Corps of Engineers will tell you that a "ponton" is a standard type of boat for bridge building and that a "pontoon" is any kind of boat used for bridge building. The distinction seems, however, to be artificial. The original in American was "pontoon" from the British, who added the extra "o" to the French "ponton." Inasmuch as the man who operated one of these boats was, in both French and in English, called a "pontonier" it was natural that the general levelling process should cause a reversion to the shorter spelling. For many decades, the army engineers used "ponton" before the rest of the army followed their lead.

POSITION OF A SOLDIER. Officially to stand at attention, but satirically to do "bunk fatigue."

PROP. In the Air Corps always speak of a plane's propeller as a "prop."

PUDDLE JUMPER. Slow airplane used for observation of limited areas and for message carrying between headquarters, and for checking positions of front-line troops. Its low flight gives it this name, but it is also called a "grasshopper," and this more widely.

PULL A GUARD. To perform guard duty, usually to "pull a good guard." Used also concerning other duties.

RAIN ROOM. An occasional description of the bath house where, according to the old song, "water runs in through a hole in the ceiling, runs right out through a hole in the floor."

RAT RACE. A mounted review in the Armored Force.

REFUGEES. Another nickname for recruits, but not much used.

REGIMENTAL MONKEY. The drum major of the band.

RIDING DOWN. Parachutist's phrase for the trip between his plane and the ground.

RIDING THE SICK REPORT. Pretending sick frequently.

ROLL UP YOUR FLAPS. Stop talking. Sporadically reappears and dies in the army.

ROOP. Slowly replacing R-SOP.

R-SOP. In the Artillery a familiar abbreviation for a combined Operation for Reconnaissance, Selection, and Occupation of a Position, which is being more freely applied to any reconnaissance, or to any "looking over" even elsewhere.

SEARS ROEBUCK LIEUTENANT. A revival, leans heavily on his uniform.

SEE THE CHAPLAIN. Stop growling. Growing rapidly in use.

SHACK MAN. A married soldier living off the army post, presumably in a shack.

SHAVETAIL COLLEGE. Training courses to prepare and select soldiers to be second lieutenants.

SKY PILOT. For the chaplain, taken of course from civilian slang of long standing, but used principally in the Air Corps.

SLAP-HAPPY. Careless and carefree as on the outside, but common in the army.

SNAFU. In limited, recent use to indicate the confusion that comes with sudden changes in orders, the word being made from the initial letters of "Situation Normal: All (Foozled) Up."

SOFT MONEY. The opposite of "hard money" or rather paper money.

S.O.L. Out of luck. World War but seems about to return.

S.O.P. Officer's slang for "Standing Operating Procedure," a command device said to have been invented by General McNair to simplify field orders and speed their execution. With the P.I.D. at Fort Sam Houston he ordered certain routine procedures for field operations and saved time and typing. It was adopted at the Leavenworth School and there written into the Field Service Regulations.

SOUP. Rain, or fog, or even clouds, as used in the Air Corps, and by them taken from civilian flying circles.

SPECIAL DUTY. A "shop" term sometimes freely used, indicates a soldier's temporary assignment as clerk, typist, PX assistant, etc. etc.

SWEAT. Especially in the air corps this is a synonym for wait. You sweat a man out when you are waiting for him. You "sweat out" a chow line while waiting for your turn for the sergeant to put your food in the mess kit.

TAILOR MADE'S. Cigarettes, not rolled by the smoker himself.

TAKE OFF. In the Air Corps this means to leave.

TURN 'EM OVER. Start the engines, used in motor units only.

TURNED IN. Reported for delinquencies.

UPSTAIRS. Used by the Air Forces for upper levels of the air. When they climb in flight they are going "upstairs."

WASHED OUT. At air fields and gaining in popularity elsewhere to mean failed to qualify.

ZEBRAS. A new word for N.C.O.'s, from their sleeve stripes.

ACKNOWLEDGMENTS

For courtesy in permitting quotations, the author is greatly indebted to the authors and publishers shown in the following list. The entry under which the quotation appears is shown as the first word in each citation.

BILLET: Arthur Guy Empey quotation from *Over the Top*, with the permission of G. P. Putnam's Sons.

BUCK, PASSING THE: William Hazlett Upson quotation from *Me and Henry and the Artillery* with the author's permission and that of G. P. Putnam's Sons.

CAISSON: E. L. Gruber's "Caisson Song," copyright, 1921, by Egner & Mayer. By permission of Shapiro, Bernstein & Co., copyright owners.

COFFEE COOLING: Morris Schaff quotation from *The Battle of the Wilderness* by permission of Sarah Schaff Carleton.

DECISION: Robert van Gelder quotation by permission of the New York *Times*.

DOG ROBBER: Westbrook Pegler quotation with his permission.

K.P.: *Yip-Yip-Yaphank* quotation with the permission of Irving Berlin, Inc.

LANCE CORPORAL: Fraser and Gibbons quotation from *Soldier and Sailor Words and Phrases* with the permission of E. P. Dutton & Co.

MEAT WAGON: "We joined the Navy to See the World" quotations with the permission of Irving Berlin, Inc.

MESS CALL: Frazier Hunt quotation, by permission, from his *Blown in by the Draft*, copyright, 1918, by Doubleday, Doran & Co., Inc.

MONKEY: Gilbert Seldes quotation from "Back from Utopia" with the permission of *The Saturday Evening Post*, copyright, 1929.

MUSTER: Oliver L. Spaulding quotation from his *The United States Army in War and Peace* with the permission of the author and of G. P. Putnam's Sons.

[233]

1314